Invented in Scotland

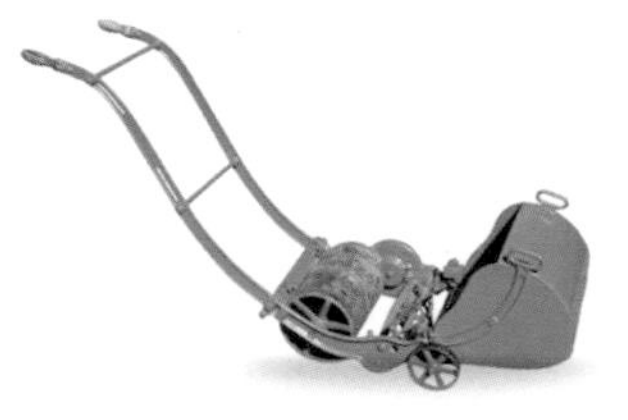

THE NATIONAL BANK OF SCOTLAND
ONE POUND

Invented in Scotland

Allan Burnett

BIRLINN

First published in 2010 by
Birlinn Limited
West Newington House
10 Newington Road
Edinburgh
EH9 1QS

www.birlinn.co.uk

ISBN: 978 1 84158 566 6

British Library Cataloguing-in-Publication Data
A catalogue record for this book is available from the British Library

Design: Mark Blackadder

Printed and bound by
Bell & Bain Ltd, Great Britain

Picture credits

Licensed through *www.scran.ac.uk*: p.12 © Historic Scotland; p.13 © Highland Council (Wick Town Hall); pp.14, 16 © University of Strathclyde; pp. 19, 40, 41, 45, 75, 88, 106, 108 © National Museums Scotland; p.20 © Douglas MacKenzie; p.21 © Scottish Life Archive; p.22 © University of Glasgow; p.48, p.58 © Newsquest (Herald & Times); p.52 © Hunterian Museum and Art Gallery, University of Glasgow; p.58 © Scottish Media Newspapers Ltd; p.62 © University of Dundee; p.77 © British Waterways; p.81 © Royal Commission on the Ancient and Historical Monuments of Scotland; p.82 © Dumfries and Galloway Council, Nithsdale Museums; p.83 © St Andrews University Library; pp.84, 85 © East Lothian Museums Service; p.104 © W. Roderick Stewart. Science and Society Picture Library, p.10; National Museums of Scotland, p.74; Imperial War Museum, p.99.

For Hannah,
invented in Scotland

CONTENTS

Introduction 7

1. In the home 11
2. On the road 27
3. In business 34
4. In the classroom 46
5. In sport 55
6. At the bank 59
7. In the hospital 62
8. On holiday 75
9. On the farm 87
10. At the crime scene 91
11. In war 96

Further reading 112

Dial it.

Tune it.

Scan it.

Clone it.

Pump it.

Float it.

Bridge it.

Blast it.

Hypnotise it.

Fingerprint it.

Shoot it.

Cure it.

Write it.

Translate it.

Stereotype it.

Post it.

Drain it.

Mow it.

Thread it.

Waterproof it.

Reap it.

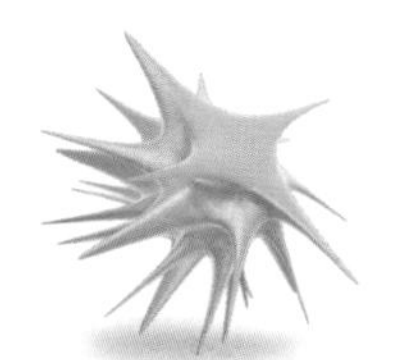

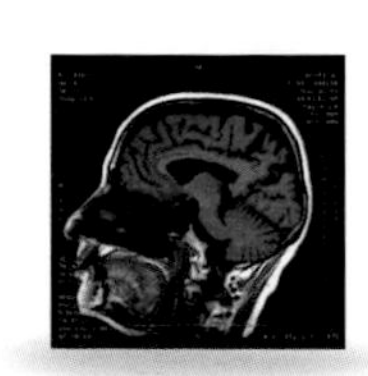

INTRODUCTION

Dilute it.

Microwave it.

Time it.

Light it.

Calculate it.

Heat it.

Cool it!

We couldn't do any of these things properly without Scottish inventions. All over the world, people's lives depend on machines, gadgets, systems and concepts that were invented in Scotland. When someone is rushed to hospital in an ambulance for a life-saving operation, every step of that journey depends on a Scottish invention. From the telephone that made the emergency call, to the vehicle's tyres, engine and the road surface; from the apparatus used to diagnose the illness, to the medicines and techniques used to treat it – all were invented, either wholly or in part, in Scotland.

This book tells the amazing story behind such inventions, how they were made and who came up with them. It explains how a fascinating and diverse group of Scottish innovators, entrepreneurs, geniuses and eccentrics built the modern world. And it reveals how great ideas and theories were turned into real practical innovations that revolutionised our everyday lives. These inventions propelled humanity out of fumbling darkness and into a brighter future by allowing us to work faster, build better, travel further and live longer.

The lives of Scotland's inventors have been by turns awesome, shocking, sad and hilarious. Their ambitions were achieved sometimes against huge odds and in circumstances of immense personal hardship. Often the competition in their field was fierce, even dangerous.

Learning about these people – men and women – is a thrilling journey of discovery. It provides an insight into why a small country such as Scotland produced such an unusually large number of inventions when compared with other, often much bigger, nations.

The story begins centuries ago, when the rest of the world considered Scotland to be a small and backward society on Europe's dark, northern fringe. Even then, the first Scottish inventors were emerging. Take late-medieval mathematician John Napier, for example. The aim of his studies was to predict the date of the Apocalypse, or Judgement Day. Strange as it may seem, Napier looked forward to this day because, according to his religious views, it was a time when the good people of the world would be allowed to go to Heaven. To make sure the Scots had time to prepare, Napier invented some simple but ingenious computing devices to try to predict the exact date. Napier's religious forecast turned out to be wrong, but his calculating methods were sound. His inventions were the forerunner of modern computers, and their principles are still at work in your laptop and mobile phone today.

From Napier's root, the numerical brilliance of Scottish inventions grew exponentially. By the late twentieth century, the list of such inventions included the first automated telling machine, or ATM, otherwise known as that hole in the wall which gives you cash. This was followed by the PIN number, otherwise known as the security code that stops your cash being given to somebody else. In the centuries in between, Scottish inventors did the numbers on velocity, temperature and even time itself. Sir Keith Elphinstone devised the first speedometer. Lord Kelvin charted the absolute coldest point at which matter exists. Sir Sandford Fleming came up with international standard time. Then there is Lord Hutton, from whom we discovered when time as we know it began. For it was Hutton who worked out that the earth is millions of years older than the Bible suggests – paving the way for Charles Darwin's theory of evolution, and the theory that the universe began with the Big Bang.

Numbers are only a fraction of the story. Scottish inventors got their hands dirty by applying their calculations to other fields – literally. Ploughing, draining, reaping, mowing and threshing: with help from such Scottish inventions as Rev Patrick Bell's mechanical reaping machine, society became more able to feed itself reliably and abundantly. The impact of Scottish inventors on food and drink didn't stop there. From commercially produced marmalade and lime cordial to vacuum flasks, paraffin, lucifer matches and microwaves – all of these come from Scotland's cupboard of big ideas.

The Scots' unique hunger for new inventions suggests that this was a nation that truly reaped what it sowed. From the sixteenth to the nineteenth centuries, the Scots traditionally placed an investment and emphasis on education which was quite far in advance of most other countries. As Napier's endeavours suggest, Scottish education had religious origins. The Presbyterian church promoted the ideal of learning for everybody. The Catholic kings Charles II and James VII supported the elite Advocate's Library in Edinburgh, which was a great centre of learning and the forerunner of today's National Library of Scotland.

These developments were the seedbed of the eighteenth-century intellectual awakening we know today as the Scottish Enlightenment. During the Enlightenment the emphasis moved away from religion and questions about belief, or faith, and on to science and questions

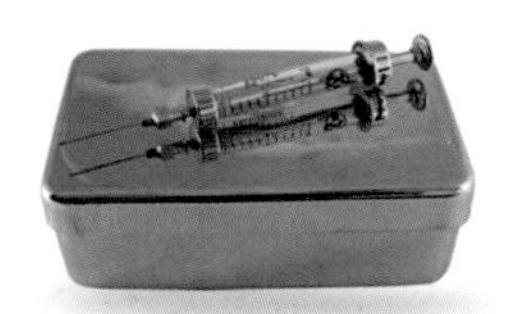

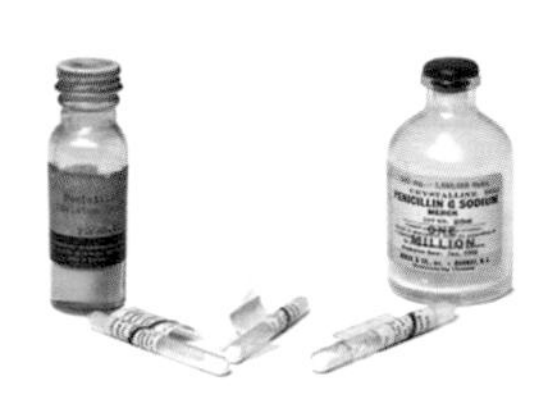

about proof, or evidence. The Enlightenment in turn nourished a century of practical scientific innovations in Scotland, from Brownian motion to latent heat and colloid chemistry. During that era, the outside world barely had time to revise its low opinion of Scotland before the Scots took that outside world by storm.

The political Union of Scotland and England laid the foundations of the British Empire, but the global success of that Empire was propelled by Scottish inventions. James Watt made steam power truly revolutionary with his crucial improvements to the steam engine. In doing so he cleared the way for Scottish inventors to fill the seas with shipbuilding firsts, including the first iron-hulled vessel. On land Scottish inventors rolled out the pneumatic tyre, the pedal bicycle and macadamised roads. They also helped create modern transport infrastructure with their innovative designs for bridges and viaducts.

If the Scottish inventors' impact on transport was big, their contribution to communications was phenomenal. The postal service, the telephone and the television are three marvels that Scottish inventors either created outright or shaped with an essential contribution. It was Scots who invented print stereotyping, the postcard and the fountain pen. And without the template for the historical novel created by Sir Walter Scott, there would likely have been no *War and Peace* and no *Tale of Two Cities*.

In medicine, the Scots prescribed new remedies at a feverish pace. For the effective treatment of scurvy, malaria, typhoid fever and tuberculosis we must thank Scottish medical pioneers, and we must do the same for the development of anaesthesia, insulin and penicillin. That's not to mention the fact that Scottish minds are responsible for clinical hypnosis, ultrasound scans and the first successful cloning of an animal.

Just as Scottish inventors have developed the means to lengthen people's lives in hospital, so too have they found ingenious ways to shorten them on the battlefield. Scottish inventors have made armies deadlier by deploying their talents on the creation of carronade cannons, rifles, gas masks and radar technology. They also helped make the police more effective against lawbreakers with Henry Faulds's development of criminal fingerprinting.

On a lighter note, the foot-pedal piano ranks among Scottish inventors' contributions to entertainment. Then, of course, there is golf. But where would golfers be in Scotland without waterproof clothing? The iconic Mackintosh raincoat took the name of its inventor, to be sewn and repaired by another Scottish first – thread from a cotton reel. Among Scotland's sartorial and musical inventions, of course, the tartan kilt and bagpipes go without saying. The list of inventions goes on, and in its brilliant diversity it is perhaps best summed up by one more Scottish creation: the kaleidoscope.

The importance of inventions to our everyday lives cannot be overstated. In fact, ground-breaking innovations are the keys to the future. But why exactly has Scotland produced such a uniquely large number of them? Why have inventors of Scottish birth, origin or citizenship been so successful? Again, part of the answer might be that once upon a time Scotland put a uniquely strong emphasis on education. Another part might be that it was the sort of place, in centuries past, where people from a wide range of backgrounds could get their ideas noticed. But in the end, this remains an open question for the enquiring and inventive mind of you, the reader, to contemplate.

ELECTRIC CLOCK
Alexander Bain, 1810–1877

The first invention to greet many of us every morning, whether we want it to or not, is the electric clock. The world's first practical electric clock was built around 160 years ago by a watchmaker from Caithness called Alexander Bain. Before Bain came along, clocks had relied on a wind-up mechanism to keep going.

All that changed when Bain, a crofter's son and bright young scholar whose first job was as an apprentice to a watchmaker in Wick, left his parents and ten siblings to move to Edinburgh and then London in 1837. While continuing to hone his practical skills as a clockmaker, Bain attended lectures on the theories and possible applications of electrical science. This gave Bain the inspiration for a clock which used electricity to drive its pendulum, and therefore would never have to be wound up.

Bain's clock, first constructed in 1840, worked by swinging a pendulum between two electric coils, through which currents were passed. The electric currents compensated for the energy lost through friction in the pendulum and clock mechanism, thereby keeping the pendulum swinging evenly and in perpetuity. The timed impulses of current in each coil required to move the pendulum between the two sides were controlled by a sliding contact, driven by the action of the pendulum itself.

After constructing some working models, Bain showed these to Charles Wheatstone, professor of physics at King's College. Wheatstone surprisingly discouraged Bain from developing his clock further, but his motives soon became apparent. Wheatstone attempted to pass Bain's innovations off as his own in a demonstration before the Royal Society of London. By then, however, Bain had already applied for a patent for his technology, which was granted in 1841, and Wheatstone was forced to withdraw his so-called invention.

Bain went on to write a book, published in 1852, called *A Short History of the Electric Clock*. In it he described his vision for a clock battery, which would provide a permanent source of power: 'If we place a sheet of zinc and another of copper in the ground a little distance from each other, and a few feet deep, so that they are embedded in moist soil, we have, by this simple arrangement, a source of electricity, and if the sheets of metal are about two square feet each we shall have amply sufficient to work a clock.'

Luckily for the rest of us, battery technology has moved on a bit since then and it is no longer necessary to dig a pit in your garden and fill it with metal plates if you want to get up for work on time. Nevertheless, Bain established important first principles, as he had done with the electric clock, for which he is duly remembered as 'the father of electrical horology'. Another of Bain's key inventions was the *fax machine* (See In Business).

Right. Alexander Bain.

Opposite. Bain's electric clock

INDOOR TOILET

The villagers of Skara Brae, c.3000 BC

Once woken by our electric alarm clock, a visit to the toilet is often required. But have you ever wondered who invented this marvellous convenience? It is a common misconception that the palace of Knossos in Crete, completed c.1400 BC, contained the world's first flushing toilet. Similarly misplaced is the competing belief that the ancient Romans introduced lavatories to the world, or that the convenience originated in India. In fact, the oldest known example of an indoor toilet, which was flushed with pots of water into a drain, is set into a wall in the neolithic village of Skara Brae on the coast of the Orkney mainland, in Scotland's Northern Isles.

Skara Brae's underground complex of domestic chambers and passageways was home to some of the earliest Scots until modes of living changed and the settlement was eventually abandoned. It became buried under the dunes until, in 1850, a great storm led to its discovery. Archaeologists later worked out that amid the box beds, stone dressers and pots, there was at least one recess in the wall that led to a crude sewer system. They had uncovered one of history's greatest inventions: the lavatory.

Above and below. One of the houses at Skara Brae.

THERMAL UNDERSOCK

John Logie Baird, 1888–1946

Time to get dressed, but make sure you wear warm, cosy socks – especially if you are in Scotland. You can walk into all sorts of clothing shops these days and buy thermal socks that keep your feet dry and comfortable through the use of various clever technologies. Did you know, however, that a thermal undersock was invented in 1914 by the Scotsman John Logie Baird? Baird, an engineer and entrepreneur from Helensburgh, suffered from poor circulation and hence cold feet. This condition inspired him to come up with undersocks which, worn under regular socks, kept the wearer warm in winter and cool in summer.

So how did the Baird undersock work? Baird realised that if he sprinkled a thin cotton sock with a mineral compound called borax, the mineral's water-absorbing and anti-bacterial properties helped keep feet dry and fresh.

The Baird sock was a profitable commercial venture. Baird's astute marketing, typified by the slogan 'Keep the soldiers' feet in perfect health', successfully targeted the British armed forces fighting in cold and wet conditions in the trenches of the First World War.

Unfortunately, the inventor's ill health forced a halt in production in 1919. Baird also invented the world's first working *Television* (see page 14).

COTTON SEWING THREAD
Patrick Clark, c.1800

Got a hole in your undersocks? No problem. Just reach for the reel of cotton thread that was invented two centuries ago by a textile manufacturer from Paisley, Renfrewshire, called Patrick Clark.

In a way, this Scottish innovation owed its real inspiration to Napoleon Bonaparte. It was Napoleon's naval blockade of Great Britain in 1806, an act of revenge following the French leader's defeat at the Battle of Trafalgar, which caused the UK's supply of cheap Oriental silk thread, the standard means of sewing fabrics together, to be cut off.

Recognising that an alternative was needed, Clark, whose silk-thread business was directly threatened by the crisis, devised a method of twisting cotton yarns together to make a long, strong and smooth thread – an ideal replacement for silk.

The Clark family business opened its first cotton-thread manufactury in 1812 and quickly flourished as cotton textiles were used to clothe the subjects of the triumphantly expanding British Empire. In 1952, the Clark company merged with another great Renfrewshire cotton business, J & P Coats, to form the textile giant Coats & Clark, now also known as Coats Plc, which employs around 25,000 people worldwide.

TELEVISION
John Logie Baird, 1888–1946

Whether it is for checking the news, weather and traffic first thing in the morning, or for vegging out on the couch after a hard day's work, the television has provided humanity with a service so complete it is hard for many of us to imagine life without it. In fact, TV was probably the defining invention of the last century. It ranks alongside the wheel and the printed word in historical importance. Without television most

John Logie Baird with his high-definition colour television.

of us would never have seen how mushroom clouds rise in the wake of an atomic blast. We would never have watched polar bears or tropical lizards hunting in the wild. And we would never have witnessed people walking on the Moon.

By beaming a virtual world into our living rooms, television takes us to places our ancestors never imagined seeing. It educates, entertains, informs and sometimes misinforms us in ways that are profound, complex and thrilling. Yet, for all that, TV's basic premise is incredibly simple: it's just a screen showing moving pictures.

In many ways television is an ancient idea. That the term tele-vision derives from classical Greek and Latin, meaning 'seeing at a distance', tells its own story. Plato was fascinated by the shadows of marionettes, lit from behind by a fire, dancing on the wall of a cave. The wall acted as a screen onto which the puppet show was projected. The principle of a screen showing moving images remains unchanged today – even in the age of ultra-thin LCD technology and internet broadcasting.

But shadows on a cave wall do not a TV make. The highly-prized title of 'inventor of television' rests not with any ancient philosopher. Instead it belongs to a Scottish jam manufacturer called John Logie Baird.

The son of a Kirk minister, Baird was born in West Argyle Street, Helensburgh, on 13 August 1888. As a boy, Baird, the youngest of four children, showed an aptitude for things electrical and mechanical. After finishing school, in 1906, he was enrolled at the Glasgow and West of Scotland Technical College to study electrical engineering. Baird was an impressive student and graduated to a job as assistant mains engineer with the Clyde Valley Electrical Power Company.

In 1918, Baird, who was refused military service in the First World War because of chronic ill health, resigned from the Clyde Valley Company. His exit was hastened when he attempted to make artificial diamonds by passing a huge current through a pot of concrete. He succeeded only in cutting off the local electricity supply and enraging his superiors. However, the spirit of invention and entrepreneurialism shown by this experiment was what propelled Baird on to greater things.

By this time he had already devised his Baird Undersock (see above). During stints in Glasgow, London and the West Indies he came up with further innovative ideas for commodities and ways of selling them, including jam, boot polish, fertilizer and soap. By the early 1920s, however, his chronic ill health was getting the better of him, and he retired to Hastings to recuperate. It was there that he applied his inventive mind and his mechanical and electronic knowledge to the question of transmitting and receiving moving images. In other words, he began drawing up plans for a television.

Without much money to his name, Baird built his first primitive mechanical 'television' in 1924 using components that included a washstand, a tea chest and a biscuit tin held together by string and sealing wax. He gave what is believed to have been the world's first TV demonstration in front of an audience of fifty members of the Royal Academy of Science, in his attic workshop in 1926. The apparatus

John Logie Baird with his Telechrome, the world's first cathode ray tube for colour television.

was limited to projecting crude images at a resolution of barely 30 lines.

To begin with, Baird could only transmit visual signals across a few feet. By 1927, however, he was able to transmit television along 438 miles of telephone line between Glasgow and London. The following year, his Baird Television Development Company successfully transmitted TV across the Atlantic.

Baird's innovation was marvellous for its time, and he even used it to make primitive video recordings that still survive, but it was not without serious limitations. Fear of an embarrassing public failure caused by a technical hitch saw Baird refuse to take up a £1000 bet set by *Popular Wireless* magazine in 1928 for him to transmit pictures of simple objects in motion, such as marbles and the moving hands of a clock, over a distance of 25 yards.

What's more, Baird also had to face down rivals for his crown as inventor of TV. The development of the technology needed to make television possible actually began before Baird was born. At the heart of Baird's mechanical 'television' – a term coined by Constantin Perskyi at the first International Conference of Electricity in 1900 – was a rotating metal disk technology, which had been demonstrated in 1884 by a German scientist called Paul Nipkow. Along with the piecemeal discoveries of a dozen and more scientists, Nipkow's 'electric telescope', which scanned moving images with its large disk, was an essential component of Baird's TV.

A major problem with the Nipkow disk was that while the machinery was large, the viewable image was small. This was not so much of an issue for the pioneers of an entirely different, rival technology – the electronic television. While Baird got his mechanical system into the public domain before any other type of working TV, by the time his experimental television service was begun by the British Broadcasting Corporation in 1929, the technology behind it was already obsolete.

The type of television that would go on to conquer the world's living rooms was not the large and cumbersome TV pioneered by Baird, with its huge disks and small viewable image. Rather it was the electronic system based around a device called the cathode ray tube, which offered higher resolution and a much larger viewable image, and this approach was being pursued by others as far and wide as Russia and the US.

Like characters from a science-fiction serial, the pioneers of electronic television bore such exotic names as Boris Rosing, Vladimir Zworykin and Philo T. Farnsworth. Unlike Baird, many of them were very well financed and supported in their research. Using a combination of Nipkow's disk and a cathode ray tube, it was Rosing who in 1906 built a prototype mechanical-electronic TV system. Rosing and Englishman A. A. Campbell-Swinton then developed purely electronic scanning methods of reproducing images. Russia's Zworykin, who pioneered the modern television camera, and American teen-genius Farnsworth, advanced this electronic model.

It was left to Guglielmo Marconi, an Italian, who shared Baird's entrepreneurial flair and inventive spirit, to establish an American company that would develop the electronic system – and destroy Baird TV. In 1935 a BBC committee examined a side-by-side trial of Marconi's all-electronic television against Baird's mechanical device. The superior picture quality of the electronic system put its future deployment beyond question. With just a few hundred mechanical television sets in use worldwide, Baird was dropped.

Baird picked himself back up again and kept on inventing. In response to the limitations of his earlier machines, he went on to build televisions with huge screens and colour images, as well as demonstrating three-dimensional television – innovations that enhanced his contribution to the development of television as we know it. Baird carried on working through ill health right up to his death, from a heart attack, at his rented home in Sussex on 14 June 1946. He was buried in his native Helensburgh.

Being the first to invent a working television had been the dream of many ever since Faraday and Clerk-Maxwell revealed the possibilities of electromagnetic communication in the 1800s, possibly even since the time of Plato. No single individual turned that dream into reality, but Baird, a brilliant and prolific inventor working on a shoestring budget, remains the first person in history to demonstrate publicly a working TV system. For that reason, if the title of inventor of television can be claimed by one man alone, then that man is John Logie Baird.

Baird also invented the *Thermal Undersock* (see page 12).

REFRIGERATOR
William Cullen, 1710–1790

Milk, cheese, meat, fish, salad, fruit juice . . . how would we keep any of them fresh without a fridge? More to the point, who invented the refrigerator? The answer is, of course, a Scot by the name of William Cullen. Born in Hamilton, Lanarkshire, in 1710, Cullen served as a surgeon on board a merchant vessel in the West Indies, before rising through the ranks of medicine and science to become professor of medicine and chemistry at the University of Edinburgh.

It was in 1748, while working in Edinburgh, then the centre of the ideological and practical innovation we now call the Scottish Enlightenment, that Cullen gave the first known demonstration of artificial refrigeration. His system consisted of a pump, which he used to suck air out of a bell jar and create a partial vacuum over a container of diethyl ether, a colourless liquid with a low boiling point.

As the depressurisation of the liquid caused it to boil, it absorbed heat from the surrounding air, which became so cool a small amount of ice was produced. Had he chosen to, Cullen could have used this technology for keeping foodstuffs cold, but it was thought the demonstration had no practical purpose at the time.

The principle was taken up in the following century by a variety of scientists, with African-Americans Thomas Elkins and John Standard patenting refrigerator designs in the 1870s and 1890s respectively. As for Cullen, in his lifetime he, like the rest of the world, stuck to traditional methods of keeping food cool, such as packing winter ice into underground chambers. Yet, Cullen, one of the founders of the Royal Medical Society, was the first person to demonstrate, albeit inadvertently, a machine that could keep food and drink cool – the forerunner of the fridge in your kitchen.

MICROWAVE, CELLPHONE, WIRELESS AND COLOUR PHOTOGRAPHY
James Clerk Maxwell, 1831–1879

The microwave oven. The radio. The mobile phone. The wireless home network. The colour photograph. All of these essential technologies owe their existence to the pioneering work of Edinburgh-born scientist James Clerk Maxwell. Arguably the greatest physicist of all time, whose work foreshadowed Einstein, Maxwell revealed the nature of electromagnetic radiation and invented a scientific discourse to explain it in his major work *A Treatise on Electricity and Magnetism*, published in 1873.

Maxwell, a son of the gentry who grew up on his father's estate in Kirkcudbright, where he later did much of his writing, theorised that all electric and magnetic energy is emitted in waves, and that these waves travel at a velocity equivalent to the speed of light, around 300,000 kilometres per second. This brilliant insight made it possible to deduce that light itself is only one form of electromagnetic radiation, all forms of which are identical but for the differing wavelengths and frequencies that give them their individual characteristics. Not bad for a man whose childhood eccentricities at Edinburgh Academy earned him the nickname 'Dafty'.

As other scientists built on Maxwell's work, the full electromagnetic spectrum was discovered. It ranges from long waves emitted at a low frequency – such as electric waves and radio waves – through to shorter waves with a higher frequency, such as microwaves and infrared radiation, which generate heat. The shortest wavelengths and highest frequencies range from visible light and ultraviolet radiation through to X-rays and gamma rays.

While radios, mobile phones and wireless routers use waves to transmit information, microwave ovens emit microwaves to generate heat. Microwaves carry a huge amount of electro-magnetic energy that agitates the molecules of the porridge, scrambled eggs, chicken or vegetables inside the oven, causing friction as the molecules rub against each other. That friction, like rubbing your hands together, generates the heat that cooks your meal.

Maxwell, whose natural talent and hard work were far more significant factors than his father's good connections in giving him a glittering academic career at Edinburgh, Cambridge, King's College London and Aberdeen's Marischal

College, also made seminal contributions to our understanding of gas, colour and optics, including profound insights into the hitherto mysterious question of how the retina of the human eye works.

A sense of humour and patriotic pride were also revealed when Maxwell made the world's first demonstration of colour photography using a tartan ribbon. In 1861, for an audience at the Royal Institution in London, Maxwell arranged for a professional photographer to take three black-and-white photographs of the same tartan ribbon. The slides were then projected through three filters – red, green and blue – one colour for each slide. When the three were all projected at the same spot on a screen, a faithful full-colour image of the original ribbon could be seen.

Perhaps the greatest illustration of Maxwell's combined achievements, one that reflects another of his great works, a study of the rings of Saturn, is the sun that – usually – greets us in the morning. When sunlight hits our face we can literally see and feel the waves whose existence and behaviour Maxwell theorised. The sun gives us light, creates colours, makes us warm and, if we stay out in it too long, cooks our skin thanks (or no thanks) to the range of waves it emits. Were he around today Maxwell, as a man of taste and modesty, might have appreciated that sunshine is like a microwave ready-meal – something to be enjoyed in moderation.

MATCHES
Isaac Holden, 1807–1897

Many people still use gas cookers to boil their eggs in the morning or make dinner. But when the pilot light won't work – or indeed when there is a blackout – it is time to reach for a box of matches. The match, also known as the Lucifer or friction match, was invented by Isaac Holden, an entrepreneur from Hurlet, near Paisley, in the 1820s.

Sure, there had been 'matches' of a sort before that. The historical trail burns as far back as sixth-century China, when pine sticks were coated in sulphur and lit by a spark. But these volatile and cumbersome early matchsticks were not self-igniting and, moreover, were so dangerous there was a strong chance of roasting your own face off.

In Europe the age-old system of tinderbox and flint held sway until Holden, an inventive and entrepreneurial spirit who worked as a teacher, textiles manufacturer and politician, devised a recipe for modern matches. The combination of substances that caused the match to ignite when struck included antimony – a brittle, metallic crystal – potassium chlorate, gum and sulphur. A friction match was also invented independently at around the same time as Holden's by a chemist from Stockton-on-Tees called John Walker.

Holden's focus was on matters other than matches. Becoming a Liberal MP in 1865 and building a reputation as a respected authority on commerce, he did not patent his invention. Meanwhile Walker's 'Congreve match', nicknamed the Lucifer match, became commercially successful and precipitated a rise in the number of smokers – although the toxic materials involved in its manufacture meant the match was in need of further development.

The invention of Holden and Walker was the forerunner of the modern safety match, patented by Sweden's Johan Edvard Lundstrom in 1855. This was refined in 1910, when the Diamond Match Company patented the first non-toxic match in the United States, a patent that was made publicly available the following year.

MARMALADE
James Keiller, 1775–1839

A controversial topic for breakfast-table discussion: who invented marmalade? The answer to this question is two-fold. If, on the one hand, we are talking simply about commercial breakfast marmalade, then this innovation belongs to a Dundee grocer called James Keiller and his mother Janet. It was they who set up the first marmalade factory in 1797.

If, on the other hand, we are talking about marmalade more generally, then the answer is less straightforward. The earliest known use of the term 'marmalade' was in the fifteenth century and referred to a preserve that originated in Portugal, made from quinces, an astringent golden-yellow fruit. However, the ancient Greeks are thought to have preserved quinces earlier by cooking them slowly in honey, while a fifth-century Roman cookbook, *Apicius*, also gives a recipe for preserving whole quinces in honey. That said, these ancient recipes were quite far removed from marmalade as we commonly understand it today.

It is said that Mary Queen of Scots ate a 'marmalade' of quinces during the 1561 voyage that returned her to Scotland after her childhood exile in France. By this time, or soon after, quinces were giving way to citrus fruit such as oranges, which were made into a bittersweet, pulped marmalade more in keeping with what we would recognise.

This is where the Keillers come in. An apocryphal tale has it that a Spanish cargo ship stuffed with Seville oranges got stormbound in Dundee harbour and was forced to sell its consignment at a knock-down price. The buyer was James Keiller. He took the oranges home to his wife, who, in a Eureka moment, came up with a new recipe for marmalade that involved cutting the orange peel into shreds or chips and reducing the mixture down into a runny jam rather than the thick paste of earlier versions of marmalade. This meant the marmalade of Janet, a canny Scot, was greater in quantity and made more pots to the pound.

The truth is that there was no Spanish cargo ship. James, who was unmarried at the time, got his hands on oranges in an undramatic purchase, and sugar probably from the refinery near his home. He worked with Janet – actually his mother – to come up with the innovative product described above. When Keiller's marmalade became a local sensation, the further innovation of setting up a factory was followed by that of marketing Keiller's orange-shred marmalade as a breakfast spread, rather than as a dessert, which was how marmalade had traditionally been consumed. When James died, his widow, Margaret, took over the successful running of the business and Keiller's thereafter went from strength to strength.

So, taking the long and the short answer, the issue boils down, as it were, to this: if your definition of true marmalade is a bitter-sweet, jam-like breakfast spread made with oranges that contains bits of shredded or chipped peel, and is sold in jars, then as far as you are concerned there is no question about it: the Keillers invented marmalade.

An early Keillers' marmalade jar.

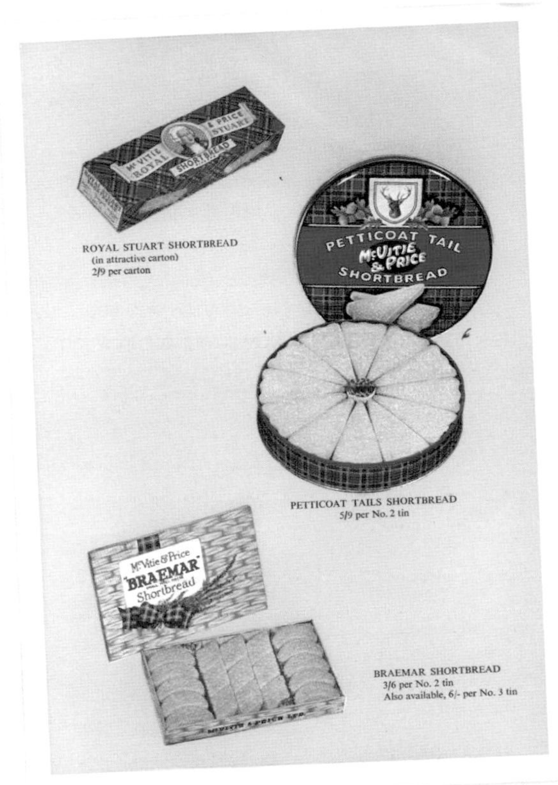

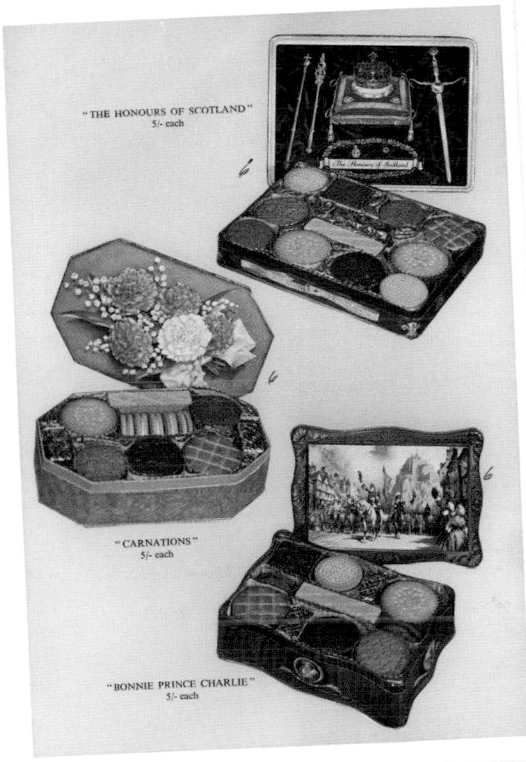

Alexander Grant added digestives to the selection of biscuits offered by McVitie's.

DIGESTIVE BISCUIT
Alexander Grant, 1864–1937

Morning snacktime comes around and you need something to go with your coffee, or to dunk in your tea. Or perhaps you want a treat before bedtime. So make it snappy. Make it sweet. Make it slightly salty. Above all, make it round.

Are we talking about cookies? Give me a break. We are, of course, referring to the king and queen of the biscuit universe: the digestive and chocolate digestive. These were invented in 1892 by a young baker from Forres called Alexander Grant, who just a few years previously had begun as a humble apprentice at McVitie's biscuit factory in Edinburgh on a wage of £1 a week.

The digestive biscuit took its name from the fact that Grant, who had a voracious appetite for literature about baking and had attended lectures on food chemistry at Edinburgh's Heriot-Watt College, claimed the high amount of baking soda in the recipe, which also contained among other ingredients wholewheat flour and sugar, acted as an aid to digestion. Whether this claim was true or not, the digestive was a huge commercial success.

Grant rose to become general manager of McVitie's in 1910 and – thanks to his profound understanding of food, his commercial nous and his robust work ethic – an extremely wealthy man. He was made Sir Alexander Grant, baronet, in 1924, ostensibly for his public services, including several large cash donations to such institutions as the National Library of Scotland.

Grant's biscuit empire almost crumbled, however, when it was suggested that he actually received his baronetcy in return for a favour to then prime minister Ramsay Macdonald, in the form of a luxury automobile and a loan of £40,000. Grant survived this scandal, which he presented as a storm in a teacup, and the following year his company unveiled the chocolate digestive – another hit.

Noted for working his employees hard, Grant remained a hard grafter himself right up to his death from a fatal bout of pneumonia in 1937. By that time Grant's digestive had become the cornerstone of a global confectionery empire, now United Biscuits. The digestive remains one of the most popular food products in the world, with around six million of them eaten every day in the UK alone.

SQUASH
Lachlan Rose, 1867

No, not the game. We are talking about juice cordial, commonly referred to as squash, the concentrated fruit drink that is diluted with water and found in fridges and cupboards the world over. The world's first concentrated juice was Rose's Lime Cordial, patented by one Lachlan Rose, who hailed from a family of ship repairers in Edinburgh's port of Leith. His innovation came about after he saw an opportunity for importing lime juice from the West Indies and making a tidy profit.

Concentrated lime juice was given to sailors on long voyages to combat scurvy, a fatal wasting disease caused by a deficiency of vitamin C. In 1867 an amendment to the Merchant Shipping Act made it compulsory for all British merchant ships out of port for 10 days or more to carry lime juice rations. To stop the juice going bad by fermenting during long voyages it was laced with rum – one gill of rum to three gills of citrus juice back in Nelson's day. But this was neither very cost effective nor ideal for sailors, who had to keep their wits about them.

Within a year of the new regulations coming into force, Rose patented a means of preserving concentrated lime juice without the need for alcohol. He worked out that the preservative sulphur dioxide would block the fermentation process. When he added sugar to the mixture, Rose also realised that the drink was eminently palatable and had the potential to be sold in bottles to a wider market. In so doing, Rose invented the world's first

Joseph Black.

branded fruit juice, Rose's Lime Juice Cordial. The company bought its own lime plantation in Dominica in 1893 and Rose's cordial drinks remain popular to this day.

FIZZY DRINKS
Joseph Black, 1728–1799

The essential ingredient in fizzy drinks, carbon dioxide, is a gas that dissolves in water under pressure and forms the bubbles that give drinks like cola and lemonade their fizz. It was discovered by one of the great scientists of the Scottish Enlightenment, Joseph Black. Born in Bordeaux to a mother from Aberdeenshire and a father from Belfast, Black lived his entire adult life in Scotland, where he distinguished himself at the universities of Edinburgh and Glasgow in the fields of physics and chemistry.

In 1754 Black conducted experiments that revealed the existence of carbon dioxide, or, as he called it, 'fixed air'. He knew this gas had to be different from normal air, because when he created an atmosphere under a bell jar containing only fixed air, neither a flame nor a mouse could breathe in it.

Black's discovery of 'fixed air' was hugely important, not just as far as the future of fizzy drinks was concerned, because it was the first time anyone had separated one gas from all the other gases that make up air. Before Black's experiment, it was assumed that the atmosphere was made up of just one element. Black showed that air was actually made of lots of different elements, one of them being carbon dioxide, or CO_2. Black, who published his findings and drew acclaim for several other theories and discoveries, such as latent heat, invented a whole new approach to chemistry with his discovery and theories about carbon dioxide.

Years later, in 1767, another scientist called Joseph Priestley developed Black's discovery and used CO_2 to create carbonated water. In the 1790s an enterprising amateur scientist from Geneva called Jacob Schweppe built a factory to produce commercial carbonated water – and the company Schweppes, a byword for fizzy drinks, was born. Mind you, as every Scot knows, the ultimate fizzy drink is arguably another Scottish invention: Iron Brew, or Irn Bru, first produced by A. G. Barr of Glasgow in 1901.

WHISKY
Traditional

A spirit that goes well with soda water and other carbonated drinks is, of course, whisky. A spirit distilled from fermented grain mash and aged in wooden barrels, whisky is as Scottish as haggis, kilts and heavy rain. The distinctive character, traditions, processes and regions associated with Scotch whisky production mean the question of whether this drink is a Scottish invention has to be a no-brainer. Or is it?

Sure, the Bushmills distillery in Ireland was licensed to produce whisky in 1608 by King James I (James VI of Scotland) and no older documents relating to any Scottish distilleries survive. But, then again, Bushmills is part of an Irish tradition producing Irish whiskey – a difference of spelling that denotes a difference of character and tradition.

What's more, because of a lack of surviving historical evidence from earlier times, nobody can say with any certainty whether the distilling of whisky – or whiskey – originated in Scotland or Ireland. The drink's history might go as far back as early medieval times when the kingdom of Dal Riata, in what is now Argyll, was a melting-pot of incipient Irish and Scottish culture. This helps explain why whisky's original name is *uisge beatha*, a term that comes from the Gaelic language common to the two countries.

Uisge beatha is a Gaelic translation of the Latin *aqua vitae*, meaning 'water of life'. Of course, *aqua vitae* is a term long used across Europe to refer to a variety of spirits, including schnapps and vodka. But the Gaelic *uisge beatha* referred to a drink specific to this part of the

world. Over the centuries the name was rolled around the tongues of speakers of Scots – the nation's other historic language, not to be confused with Gaelic – and its dialectal cousin, English, as uiskey-ba, uiskey and, finally, whisky.

It is worth pointing out that while Bushmills of Ireland became an officially registered company in 1784, producing spirits in the modern commercial sense, an advert in the *Aberdeen Journal* the following year reveals that the Glen Garrioch Distillery – now the Meldrum Distillery – in the Scottish Highlands was also up and running. Other evidence suggests that the Strathisla distillery on Speyside may have predated both of these, making it possibly the oldest commercial whisky distillery still in existence today.

LAWNMOWER
Alexander Shanks, 1801–1845

These days the lawnmower is a device so advanced some examples almost rival small cars for their power and sophistication. Not to mention that the size and style of your lawnmower can be symbolic of your status in suburbia. It is refreshing to note, however, that the humble, unmotorised mower is still on the go – a contraption that owes its origins to a machine patented by engineer Alexander Shanks in 1842.

Shanks's invention was the world's first effective lawnmower, although others had tried and failed to devise one previously. It could cut and roll grass in a single operation. The 2ft-wide mower was horse-drawn and the driver walked behind, guiding the horse using reins. Shanks, who was born in Forfarshire, had a general interest in the mechanisation of labour. His company, Alexander Shanks & Son, produced a variety of machines, including steam engines and diggers.

After Shanks's untimely death from consumption aged just 44, his son, James Shanks, took up the reins of the company and the mower, a large version of which he marketed brilliantly with a demonstration at the Empire Exhibition of 1851 in London, and later at the Paris Exhibition of Industry.

Above. Nineteenth-century lawnmower advertisements.

KALEIDOSCOPE
David Brewster, 1781–1868

Long before games consoles, action figures, frisbees, space hoppers or football cards, the imagination of children and adults was captured by the kaleidoscope. Invented in 1816 by David Brewster, the son of a schoolteacher in Jedburgh in the Scottish Borders, the kaleidoscope was devised as a tube of brass or card containing mirrors and one or more objects, such as beads or stone crystals, that were to be rotated by the user. When the kaleidoscope was turned, the reflections of the objects in the mirrors created perfectly symmetrical, vividly colourful patterns. As Brewster noted with great excitement, these were beautiful and pleasing to the eye.

A kind of kaleidoscope was supposedly used by the ancient Greeks, but it was Brewster who came up with the kaleidoscope whose basic principles remain unchanged today. He did so while conducting optical experiments, the more 'serious' side of which saw him make important theoretical and practical contributions to the emerging technology of lenses in lighthouses and cameras. Brewster was also an astute self-publicist, using journalism to compensate for his shyness when it came to public speaking. In 1831 he set up the British Association for the Advancement of Science.

Unfortunately, while Brewster might have been expected to become very rich on the back of his kaleidoscopic invention, his design was so exciting that the instrument-maker charged with constructing the prototype blabbed its secret to the world. As a result, Brewster's patent application was approved too late to prevent a rash of copy-cat kaleidoscopes from appearing in the shops. Through his various enterprises and political connections, as well as academic honours, Brewster nevertheless became a wealthy man and had a house built for his family near Melrose, also in the Borders, where he died in 1868. He was buried in the graveyard of Melrose Abbey.

SCOTTISH TERRIER
Traditional

When varmints such as rats, foxes or badgers appear on a farm and start scoffing grain, eggs or chickens, it is time to call in the cavalry. In fact, so adept are Scottish terriers at flushing out vermin with their endless determination and capacity for digging, they have earned the nickname of 'little diehards' and inspired the creation of a regiment of Scottish soldiers, the Royal Scots, whose leader was a passionate fan of the breed.

There is no doubt that Scottish terriers are a breed apart. In fact, make that five breeds – Aberdeen (often known today simply as a Scottie), Skye, Cairn, Dandie Dinmont and West Highland. These hardy little dogs, as the name suggests, originated in Caledonia, although when exactly this happened is a mystery.

Historical records suggest the Scottie may have been around in the fifteenth century or earlier, but the breed was definitely established by the late 1500s. Scotland's King James VI, who became also James I of England in 1603, was a famous fan of the Scottish terrier. By the nineteenth century several different breeds were identified, each with subtle but distinctive differences in colour, build and temperament. Developed for controlling pests in his native land's often harsh and unforgiving climate, this hardy little dog is another distinctive Scottish invention.

SPEEDOMETER
Sir George Keith Elphinstone, 1865–1941

When you leave home and need to get to work, Scottish inventions help make it happen. Take the motor-car speedometer, for example. This was invented in the 1890s by an aristocratic electrical engineer from Edinburgh called Keith Elphinstone. From an early age Elphinstone showed an aptitude for technology, something which ran in the family. His uncle, the fifteenth Lord Elphinstone, owned the Elphinstone-Vincent Electro Dynamo Machine Company – and it was here that young Elphinstone cut his teeth in the mid-1880s before going on to further his electrical education with several other companies in various locations.

In 1891, Elphinstone bought his own firm, the electrical-instrument makers Theiler & Sons, in London. Elphinstone became an industrious innovator and patented several inventions. His specialism was in electrical instruments for transport and the field of battle. He developed speed-recording equipment for trains, bombsights and other instruments for warplanes, as well as fire-control systems that helped naval gunners hit their targets during the First World War. However, his invention of a speedometer for cars, an earlier achievement than the speedometers that were devised by rival inventors in the early 1900s, is for most of us his most visible legacy.

PNEUMATIC TYRE
Robert William Thomson, 1822–1873, and John Boyd Dunlop, 1840–1921

Imagine driving an automobile, bike or truck without tyres. Ouch. It's achingly clear the pneumatic tyre was a major innovation, and credit for it is traditionally given to John Boyd Dunlop, a farmer's son from Dreghorn in Ayrshire.

After qualifying as a vet in Edinburgh and establishing himself in Belfast, Dunlop arrived at the idea of an air-filled rubber tyre in 1887 while watching his son playing on his tricycle. The trike jolted and jarred uncomfortably as the boy rode it over the paved street. The solution, Dunlop realised, was to make a rubber air tube and fasten it around the rims of the tricycle. It was a huge improvement. After a successful public demonstration of his tyres the following year, Dunlop quickly patented his invention.

Or was it actually *his* invention? Dunlop's patent was invalidated when it turned out the principle of the pneumatic tyre had been invented forty years earlier, by another Scot called Robert William Thomson.

A merchant's son from Stonehaven, Thomson was a largely self-taught civil engineer and entrepreneur. When he was just fourteen he was sent on a formative sojourn across the Atlantic to Charleston, USA, to learn about business. Once back in Scotland Thomson became an engineering apprentice, and among his early jobs was helping to demolish a castle. He came up with the idea of electrically controlled explosions and became a blasting specialist. On the back of the railway boom of the late 1830s and 40s, Thomson set up his own business as a railway engineer in England. When the railway bubble burst, Thomson went out of business. That's when he began turning his mind, which had been attuned to solving transport problems by his experience on the railways, to a new invention.

In the 1840s it was becoming obvious to some people that road travel was bumpy and uncomfortable and needed to be improved if the then-new technologies produced by the Industrial Revolution were to take off. Like the later Dunlop, Thomson came up with the idea of creating a cushion of air between a vehicle's wheels and the road. In 1845, Thomson took out a patent for a pneumatic rubber tyre or, as he called it, an 'aerial wheel'.

So why didn't Thomson – rather than Dunlop – go on to become the household name for tyres? Thomson was a victim of his own ingenuity. Bicycles were still a rare, developing technology; and it would be another forty years before Carl Friedrich Benz invented the modern motor car. Plus the costs of rubber and manufacture of Thomson's tyre were high. In short, Thomson had invented a product the market wasn't ready for.

This was a shame for a man who embodied the spirit of invention. Thomson went on to travel the world as a successful engineer and came up with a patent for a fountain pen in 1849, a new design for sugar-refining machinery and a portable steam crane among other innovations. He also revisited his rubber tyre and adapted it to create an early kind of caterpillar track, which worked very well – but again the cost of rubber counted against its commercial viability. Thomson died in Edinburgh aged just 51, and his reputation, like his tyre, quietly deflated.

Back to Dunlop. There are at least three good

reasons why he continues to deserve a share of the credit. Dunlop came up with the idea for a pneumatic tyre by himself, albeit a lot later than Thomson. Dunlop invented several bits and pieces essential to the working of an efficient tyre, such as valves and rims, which he patented. And the firm he established, which became known as the Dunlop Rubber Company, grew to be a household name that popularised pneumatic tyres throughout the world.

So it could reasonably be argued that it was not one, but two, Scots who created the pneumatic tyre to stop us all bumping over cracks and potholes. You would be forgiven for wondering what this says about the quality of Scottish roads.

THE PEDAL BICYCLE

Kirkpatrick Macmillan, 1812–1878

The success of Dunlop's re-invention of the pneumatic tyre was propelled by the success of another Scottish innovation, the pedal bicycle. By the late 1800s, the modern bicycle as we know it had taken shape and was on its way to conquering the world. The pedalling mechanism was an essential part of its success and the honour for that invention goes to a blacksmith from Thornhill, Dumfriesshire, called Kirkpatrick Macmillan.

Macmillan, one of at least eight children, followed his father into the blacksmith trade. He hit on the idea of a self-propelled, pedal bicycle while working as an apprentice on the Duke of Buccleuch's estate at Drumlanrig. Macmillan observed a hobby horse, also known as a velocipede, being ridden along the road and decided to make one for himself. The hobby horse – a wooden cycle that was propelled by the rider pushing off the ground with his feet – had a long history. The patent for the first practical, commercially successful version had been made by Baron Karl von Drais in Germany in 1818.

When Macmillan learned to ride his 'running machine', as hobby horses were also known, he realised that it would be improved massively if it could be propelled without having to be pushed along by his feet. Working in his father's smithy, Macmillan constructed a machine that echoed the principle of how a locomotive steam engine delivered power from its pistons to its wheels. Using connecting rods and cranks, Macmillan reciprocated the movement of the rider's feet on two pedals to the rear wheel. He completed his new bike in 1839, and was soon seen learning to ride it along the country roads, often for several miles at a time, which was no mean feat considering the machine was a heavy and unwieldy piece of kit by today's standards.

In the summer of 1842, Macmillan rode no fewer than seventy miles to reach Glasgow. During the journey, which took him two days, he apparently bumped into and slightly injured a girl who had run out in front of him, for which he was fined five shillings. Nevertheless, his cycle impressed a lot of people and soon copies of it were changing hands for a few pounds apiece. One enthusiast for the new design was Gavin Dalzell from Lesmahagow, which prompted the belief for many years that Dalzell was the inventor.

The trouble was, Macmillan never patented his invention, which has prompted cynics to question his place in the history of the bicycle. Some even argue that the attention-grabbing cyclist whose minor collision made it into the Glasgow press was not, in fact, Macmillan. But if not him, then who?

THE PUSHROD ENGINE
David Dunbar Buick, 1854–1929

Corvette. Mustang. Viper. Roadmaster. Legendary names on the highways of the US, and indeed throughout the world. What have they all got in common? It's the type of engine that gives each of them their distinctive grunt – the pushrod. For the non-mechanics out there, pushrods open and close overhead valves that let fuel and air into the engine and exhaust gases out. Many car engines nowadays use an arrangement called an overhead camshaft, instead of pushrods, but that type of engine is really a modification of the basic pushrod design.

So why is all of this important? There are three reasons. The first is that the invention of the pushrod engine was a major event in automotive history. Before it came along, car engines were, frankly, not up to much in the performance department. The second is that, while the pushrod is, as mentioned above, the forerunner of the overhead camshaft engine, it is also still being produced in its own right – and not just in muscle cars. Even the ubiquitous little Ford Ka has used a pushrod engine.

The third reason is the most important. The pushrod engine was invented by an engineer from Arbroath called David Dunbar Buick – yes, you guessed it, also the founder of Buick cars, which formed the cornerstone of General Motors.

As you will also have guessed, this entailed Buick migrating from Scotland at an early age to the States. Detroit – where else? – to be exact. After he left school in 1869, Buick entered an unlikely line of work, considering what he got up to later. He worked for a plumbing supplies business, and by 1884 had become co-owner. But while it seems some way removed from car engines, plumbing gave Buick the opportunity to flex his inventing muscles. His first great success was to devise a way of coating cast-iron baths with white enamel – a method that is still used today.

Being a mechanical engineer at heart, Buick grew tired of solving problems in the bathroom and began to experiment with something he was really passionate about: motor engines. He sold his plumbing business in 1902 and set up the Buick Manufacturing Company with the proceeds. The new outfit had two aims, to manufacture cars and to sell them. But Buick focused his attention on the manufacture and development side, which had two outcomes: he ran out of money and became indebted to his financiers, but before doing so developed a car that contained his revolutionary new engine concept.

Buick had developed the pushrod engine, also known as the valve-in-head or overhead valve engine, in order to solve the inherent limitations of the conventional side-valve, or flathead, engines of the day.

Left. A 1960 Corvette Convertible, powered by a pushrod engine.

Above. An early Buick Roadster.

Buick's engine was much more powerful and efficient, and went on to become the standard for automobile powerplants.

Unluckily for Buick, before he could reap the rewards, his debt-ridden company was bought out and he left in 1906 to pursue other, less successful engineering ventures. In 1908 the Buick Motor Car Company – which retained its founder's name even after his departure – was incorporated into the incipient General Motors (GM) and the rest, as they say, is history.

The overhead valve engine went from strength to strength, acquiring muscular names like 'nailhead' and 'big block'. Buick cars and their throbbing engines became legends, playing iconic roles in films, TV shows and books, including a chiller by author Stephen King. The Buick name remains very much an American icon. But that, of course, raises a question: is Buick a Scottish or an American inventor? The answer, as every good Scots-American knows, is that he was both.

MACADAMISED ROAD
John Loudon McAdam, 1756–1836

Cars get nowhere fast unless they have a decent road to travel on. Yet roads are the sort of thing we easily take for granted because they are everywhere. They guide us home, connect us to other people and even comfort us. Lost in the wilderness and don't know how to get back to the hotel? Find the highway and all will be well. Need to get to the other side of the country at the drop of a hat? The motorway or freeway is the answer.

So inventor of the modern road is a prized title indeed. Some might think this accolade belongs with the ancient Romans. Perhaps Roman roads, with their paving stones and layers of rubble, could be regarded as the foundation of the perfect road. Others might say that the apotheosis of the modern road is the freeway, or motorway, for which the Autostrada dei Laghi from Lake Como to Milan is probably the first model. But the key invention linking these two developments is the macadamised, and later tarmacked, road.

Macadamisation is a process for building roads with a smooth, firm and durable surface; much better than muddy medieval tracks or rickety Roman roads. Macadamisation is how some roads are still built today, and the process was developed by John Loudon McAdam, an engineer from Ayr whose father was a small-time laird called the Baron of Waterhead.

When McAdam's father died in 1770, the fourteen-year-old made his way in the world by crossing the Atlantic to work with his businessman uncle in New York. McAdam made a good living and married a wealthy lawyer's daughter before returning to Scotland with his wife and two children in tow. With his business wealth he bought an estate in his native Ayrshire and became a trustee of the Ayrshire turnpikes – or roads – in 1783.

McAdam's interest in road

improvement was aroused, and, after moving south to Bristol in 1812, he became a roads surveyor and wrote influential articles on the subject. Following long travels to research how existing roads were made and what could be done to improve them, he proposed that roads should be constructed from a solid base of large stones, raised above the surrounding ground, with small stones and gravel layered on top. The road should also be built with a camber, meaning it should arch slightly in a convex shape, so that rainwater would quickly drain off, keeping the road dry and stable. It was so simple and yet utterly revolutionary.

Almost overnight, McAdam became a household name, and 'macadamised' a by-word for the new roads he pioneered. Later, when motor cars took off, the technique of applying a coat of tar or other oil-based agent to the surface to prevent dust was introduced and madadam became 'tarmacadam' and, eventually, 'tarmac'.

As well as elevating roads, McAdam had also made the business of being a roads surveyor respected in society when previously that type of work had been regarded as somewhat lowly. McAdam was not only an engineer but also an organiser of roads, demonstrating how an efficient system of interconnected roads could transform whole countries. McAdam's career was not without its low points, and he had some significant business failures, but by his death, at the impressive age of eighty, in Moffat in Dumfries and Galloway, his fame had spread across the world, with most main roads in Europe and North America macadamised by the beginning of the twentieth century.

GAS LIGHTING AND VARIOUS MECHANICAL INNOVATIONS
William Murdock, 1754–1839

The warmly glowing gas street lamp, licked by tongues of freezing fog on a cold and gloomy night, is an icon of the modern city. More than just a means of illuminating the road ahead, it is a beacon of hope, comfort and civilisation; a symbol of the progress of industrial society. Even today, although sodium lamp technology now dominates our streets, the original holders of gas lamps remain in many quarters, cherished for their beauty and elegance.

This invention was devised in the 1790s by William Murdoch, later known as Murdock, a miller's son from Old Cumnock in Ayrshire. The third of seven children, Murdock was plucked from home to work in the famous Soho steam engineering works of Matthew Boulton and James Watt in Birmingham in 1777. This was probably thanks in part to good connections, since the Murdoch family lived on the estate of the celebrated diarist James Boswell, who had visited Boulton and Watt's works the previous year. But it was also down to the young Murdock's natural assets; his facility with engines and mechanical things, and imposing physical presence, being a big lad of over six feet.

Murdock quickly demonstrated his worth to Boulton and Watt, erecting engines for them in various locations around the UK and making

Right. An early advertisement for gas lighting.

Opposite. Lighting street lamps as twilight gathers.

improvements of his own to the designs – much to the chagrin of Watt, who grew jealous of Murdock's talents. As well as improving steam engines, Murdock devised an experimental steam car and created a number of machine tools. One of his greatest innovations was the sun-and-planet gear system, often wrongly credited to Watt, which Murdock invented in 1781. This system allowed the up-down motion of a steam pumping engine to be converted into rotary motion – opening up a galaxy of lucrative industrial applications.

The following year, while overseeing mining machinery works in Redruth, Cornwall, Murdock began working on the idea of gas lighting. The inflammability of coal gas, with all its devastating consequences in mining accidents, was well known. The question was, how to harness that gas safely and effectively. Murdock experimented by putting coal in an iron kettle and heated it until the coal began to give off gas. A thimble was placed over the spout of the kettle, creating a suitable opening at which the gas could be safely ignited. When he did so, Murdock was delighted by the steady, glowing flame.

He set about building a retort – a large, airtight chamber that gasified the coal – in his garden. By 1794 he had constructed a system that lit his living room. The lighting impressed Murdock's neighbour, the young Cornish locomotive engineer Richard Trevithick. Years later, Trevithick's son, Francis, wrote that: 'Those still live who saw the gas-pipes conveying gas from the retort in the little yard to near the ceiling of the room, just over the table. A hole for the pipe was made in the window.'

Murdock later returned to Birmingham, where his gas lights were seen by George Lee, a Mancunian textile manufacturer. Lee was so impressed he first lit his own home with gas before setting up a system of almost 1000 lights to illuminate the mill of Phillips & Lee in Manchester. This was the beginnings of the gas-lighting industry and Murdock was awarded a medal by the Royal Society.

Little money came to Murdock from his invention, and he continued in the service of Boulton and Watt, where his reputation as 'an incomparable engineer' was secured through his ongoing engineering innovations. He married and, although his wife tragically died from complications in childbirth, Murdock was left with two sons, William and John. By 1817, Murdock's handsome salary from Boulton and Watt had made him sufficiently wealthy to have a gentleman's mansion built, Sycamore House, in Handsworth, Birmingham.

Visitors to Sycamore House included the novelist Walter Scott. Scott was stunned by the amazing array of never-before-seen gadgets. In place of a knocker there was a doorbell, which was operated by compressed air. The place was lit, naturally, by piped gaslight. Scott was so impressed he had both the doorbell and the gas lighting installed at his own residence, Abbotsford House, near Melrose.

In 1830, Murdock was forcibly retired by his employers, who were fed up with forking out his high wages. But Murdock continued working on new ideas and innovations, including plans to harness tidal power. He died at Sycamore House, under the glow of his gas lamps, in November 1839.

TELEPHONE
Alexander Graham Bell, 1847–1922

An office without a telephone is like a car without wheels. No business would get very far in the modern world without this remarkable invention. The telephone has been in existence for more than a hundred and thirty years, and in that time it has revolutionised the way human beings communicate. Today, mobile wireless technology means that the traditional telephone, with its chunky base and handset, long winding cable and big round dial, is now a historical artefact – although the recurring fashion for all things 'retro' means that, even today, you can still buy a new phone designed in the classic, old-fashioned, wired-up style.

Whatever the model, everyone, from granddaughters to grandfathers, technophobes to technogeeks, rich and poor, owns or uses a phone. Wherever you live and work, there will be a telephone somewhere on the premises. Telephone wires connect continents, keep us in touch with friends and family, and, when the need arises to contact the emergency services, the telephone saves our lives. Internet broadcasts, television, radio, you name it: all of these colossal innovations owes a debt to the telephone, which first transmitted human conversation across great distances.

Alexander Graham Bell.

The telephone is arguably the most important invention since the wheel. But who can justly lay claim to being its inventor? Some argue that it was a German schoolteacher called Johann Philip Reis. In 1860 Reis demonstrated a kind of toy telephone using a vibrating membrane, made from sausage skin, to reproduce sounds transmitted electrically over a wire. But the design was unreliable and the sound very poor. Reis did not take his invention seriously enough to patent it. Although a very significant contribution to telephone's development, Reis's design was, ultimately, not a practical, working telephone. Another claim for the invention of telephone rests with an American electrician called Elisha Gray. Gray was a talented and highly motivated inventor and entrepreneur, and

An early Bell phone.

founder of the Western Electric Company. However, on the very day that Gray attempted to file a patent for his telephone design in Washington in 1876, he discovered that someone else had beaten him to it by a matter of hours. That someone was a Scotsman by the name of Alexander Graham Bell.

Bell was born in Edinburgh in the early spring of 1847. At that time Queen Victoria had been on the throne for merely a decade and most people still didn't have the vote. Yet Bell's lifetime would witness not only the coming of democracy, and other huge social and political changes, but also the rise of electrically powered machines the building blocks of the technological age in which we now live. And Bell himself played a leading role in that change, by inventing a truly democratic technology. The roots of Bell's telephone come from his profound belief in the importance of communication, and his sense that the means to communicate clearly and effectively with others is a fundamental human right. Bell's father, Alexander Melville Bell, and his grandfather, Alexander Bell, were both elocutionists who developed an advanced system of 'visible speech', using special symbols, which taught deaf people how to speak. The young Bell, whose mother was deaf, followed in his father and grandfather's footsteps by first developing his talents at an academy in Elgin, where he was a pupil-teacher, and at university in Edinburgh, before becoming a speech therapist and teacher of deaf people.

Bell became fascinated by humans' vocal and hearing anatomy, and in the artificial manipulation of sound. In 1863 his father took him to see a 'mechanical man', which was able to articulate a few words, and was based on principles described in a book by a German called Baron Wolfgang von Kempelen. Aleck, as Bell was nicknamed, got his hands on a copy of the Baron's book and enlisted his older brother Melville, or Melly, to help him build his own mechanical head, which used a bellows to pump air through a larynx and lips to 'say' a few words. Bell also experimented on the family dog, a West Highland Terrier, by teaching it to growl continuously while he reached into its mouth and toggled its vocal chords and lips to produce a string of sounds that resembled the sentence: 'How are you, mama?'

The 'talking dog' was both the party piece of a young man with a sense of humour, and a serious experiment by a radical, free-thinking inventor. Bell became interested in how tuning forks artifically produce sound through vibration and resonance, the same principle at work in human vocal chords. He began to realise that if tuning forks, or another artificial material, could be made to resonate in a controlled manner by the application of electric current, then vowel sounds, consonants, and even whole sentences, could be captured and reproduced by an electrical device. He discovered that another German, Hermann von Helmholtz, had already done basic experiments along these lines, and written a book about his work, which Bell pain-stakingly translated from the original German. Bell continued experimenting with sound and electricity and was enthused by his progress.

Tragedy struck Bell twice in rapid succession, in 1867 and 1870, when both of his siblings, younger brother Ted, and later his older brother Melly, died of tuberculosis. When it looked like the remaining son was also at risk of succumbing to the disease, the family decided to up sticks from London and cross the Atlantic in the hope that Bell, now in his early 20s, would benefit from the cleaner air of the 'New World' in Canada. The Bells settled in Brantwood, a small town near Niagara Falls.

In 1871, Bell's father sent him to Boston to demonstrate the family's visible speech system at a school for deaf children. By 1873, Bell's success as a teacher earned him a professorship at Boston University. His work was so impressive that he was able to charge wealthy movers and shakers for teaching their children privately in his Boston lodgings. Among those wealthy parents were certain businessmen and professionals who would become his friends and financial backers. The two most important of these were Thomas Sanders, a leather dealer, and Gardiner Greene Hubbard, a lawyer whose teenage daughter Mabel, deaf through having contracted Scarlet Fever as a child, became Bell's pupil, his lover, and, ultimately, his wife of forty-five years.

Bell's income as a teacher funded his continuing experiments. In 1874, using the ear of a dead man, he built a machine that responded to sound, called a phonautograph. The phonautograph 'listened' to deaf pupils talking and then recorded what they said as written patterns. These patterns were used to demonstrate to deaf pupils how well their learned pronunciation of words matched the 'correct' pattern made by a person with good hearing. Bell wondered whether he could replicate the resonating membrane of the human ear to create an artificial membrane, which would resonate in response to the air waves made by sound and then turn those resonances into an electric current – a current that could be transmitted down a wire to a receiver which would then reproduce the sound at the other end. The idea of the telephone was taking shape.

Hubbard and Sanders gave Bell money and business assistance in return for a share of any profits he made from his inventions. This gave Bell enough cash to hire an assistant technician, Thomas Watson. By 1875, Bell had sent all of his pupils away, except Sanders's son, in order to focus on developing his invention. Initially, Bell had set his sights on perfecting a device called the harmonic telegraph, which was able to transmit several morse-code telegraph messages simultaneously – thereby making the existing technology of the telegraph much more efficient and cost-effective. The harmonic telegraph used several metal reeds, tuned to different frequencies, to reproduce individual morse codes as distinct patterns of sound. Hubbard and Sanders both wanted Bell to focus on getting this machine into production in order to make money, and grew impatient at Bell's ongoing attempts to create a version of the harmonic telegraph that eschewed morse code in favour of directly transmitting the human voice.

The breakthrough came on 2 June 1875. Bell and Watson were conducting a sound experiment with their prototype harmonic telegraph. When one of the reeds became stuck, Watson plucked it in order to free it. The reed produced overtones, which Bell could hear at his end. The Scotsman realised that he and Watson had stumbled upon a two-way system of turning sound into electrical current and back into sound again. If it could be done with reeds, then it could be done with the human voice. It was something of an accident, but given all the hard work and searching enquiry that had brought them this far, it was a prime example of Cervantes' dictum that diligence is the mother of good fortune. Bell had established the basis of a primitive telephone.

Bell and Watson quickly set about creating a new machine called a Gallows, which was able to transmit voice-like sounds, but not yet clear speech. This was followed by an apparatus that used a diaphragm, based on the diaphragm or membrane in the human ear, to vibrate in response to sound. That vibration was transmitted to a needle which in turn vibrated in a container of water. Speech patterns caused the needle to vibrate to varying degrees, and this in turn varied the resistance in the machine's electrical circuit. These varying electrical impulses were transmitted along a wire and reproduced at the other end, using the same principles. When Bell spoke into the machine with the words: 'Mr Watson, come here, I want to see you,' Watson, who was listening at the other end in another room, could hear him clearly. Bell had created a working model of a telephone. He drew up a patent, which was granted on 7 March 1876.

The story did not end there, for even though Bell went on to demonstrate that his telephone could work over distances of hundreds of miles, many influential business people initially regarded the telephone as nothing more than a toy. Bell had invented a device that could change the world, but was anyone actually going to use it? The answer came in May 1877, when the first telephone system was leased for business use. It helped greatly that Bell had been invited to demonstrate his telephone as part of the centenary celebrations of America's declaration of independence. The telephone grabbed the global headlines and it took off, with Queen

Victoria requesting a private demonstration and subsequently describing it as 'most extraordinary'.

The success of the Bell Telephone Company made Bell and his backers very wealthy. Several improvements were made to the technology so that users no longer had to shout as had been the case in the beginning, and the company ultimately fought off a barrage of long-running legal challenges from rival inventors and their financial and political backers, who claimed the Bell patent as their own. In 1915, Bell and Watson, by then old friends, made the first transcontinental telephone call, between New York and San Francisco, 3400 miles apart. The telephone was here to stay.

There is much more to Alexander Graham Bell's inventing genius than just the telephone. He was a true pioneer of the great age of invention that was the late nineteenth and early twentieth centuries. In his later years, at his Cape Breton Island family retreat of Beinn Breagh, which is Scots Gaelic for 'beautiful mountain', Bell pursued his interest in a wide range of fields, most notably aeroplanes and hydrofoil ships. Bell experimented with motor-powered aircraft before the Wright brothers made their famous flight, and was the key founding member of the Aerial Experiment Association in 1907, whose Silver Dart aircraft was watched by Bell as it made the first manned flight over Canada. Later, with the help of another assistant, Frederick 'Casey' Baldwin, Bell designed a hydrofoil boat, the Bell HD-4, which set a world marine speed record of 70.9 miles per hour in 1919.

Bell also invented a primitive metal detector, a breathing aid that was a forerunner of the iron lung, a rudimentary form of air conditioning which cooled his house by using fans to blow air across blocks of ice, and composting toilets. Before he died at Beinn Breagh in 1922, at the age of 75, he had spoken in an interview of his interest in the concept of solar panels to generate heat. Besides the telephone, one of Bell's greatest legacies is the unit for measuring sound, which was named after him by Bell Labs, and is known as the bel, or decibel.

FAX MACHINE
Alexander Bain, 1811–1877

Another Scot who contributed hugely to the development of communications technology was Caithness crofter's son Alexander Bain. Years before Alexander Graham Bell was born, Bain invented a type of telegraph that used electrically charged chemical paper to send messages much faster than existing, mechanical telegraph technology. However, Samuel Morse, the inventor of the conventional telegraph, argued that Bain's chemical telegraph patent of 1846 was an infringement of Morse's own patent and successfully blocked Bain's system from catching on.

Bain was, nevertheless, a highly successful innovator. He is discussed elsewhere in this book as inventor of the *electric clock*. Perhaps his most ingenious invention, however, was his primitive facsimile, or fax, machine. Applying the technology he was developing in his electric clocks and chemical telegraph, Bain attached a stylus to the end of a pendulum, which swept across a metal surface, or 'document', imprinted with information in the form of a varying sequence of dark and light spots. Each dark spot passed over by the stylus caused a variation in an electrical current passing through the apparatus. This varying current was transmitted down a telegraph wire and outputted by a synchronised pendulum on the receiving end. Every time the receiving apparatus received a fluctuation in the current caused by a dark spot, the pendulum

transmitted that fluctuation onto a piece of chemically treated paper tape, resulting in a matching sequence of spots on the tape. In this way, the 'document' was 'faxed'.

FOUNTAIN PEN
Robert William Thomson, 1822–1873

Whether for signing faxes or scribbling notes from telephone conversations, the fountain pen has a long and illustrious history. Today, ballpoint pens rule, but it was the fountain pen that first superseded the laborious quill and inkpot by offering a writing instrument that carried its own ink supply within its casing. All subsequent pen designs followed the same basic principle, making this yet another massive invention. The first fountain pens of a sort began appearing in the early eighteenth century, if not before. The only trouble was, they were rubbish. Constant spillages, blockages and huge, uncontrollable splurges of ink meant these early attempts lacked any credibility.

It was not until the nineteenth century that a reliable, clean fountain pen was designed. Credit for this innovation is often attributed to Lewis Waterman, who claimed to have invented the first 'practical' fountain pen in the 1880s. However, he was beaten to it by Stonehaven's Robert William Thomson. Noted elsewhere for his role in inventing rubber *pneumatic tyres*, Thomson constructed a self-filling fountain pen that was made of glass and which he patented in 1849. Two years later Thomson demonstrated his pen publicly to an impressed audience at the Great Exhibition.

COMPUTING
John Napier, 1550–1617

The computer is at the heart of the twenty-first-century office. As there seems to be no limit to what this technology can achieve, many of us assume it is bafflingly complex. Break down how a computer works, however, and we see that its basic building blocks are really quite simple. Every 'thought' a computer has is a straightforward calculation. By performing millions of simple calculations in an instant – a celerity made possible by electronics – a computer can perform complex tasks. The first electronic computers were developed in the 1930s and 1940s, the most significant being the Electronic Numerical Integrator and Computer, or ENIAC, of 1946, which was designed for the US Army. However, the basic principles of computing, or calculating, which allow machines to solve problems and process data using a mathematical 'program', were established long before that.

The foundations of modern computing were laid around 1600 by a wealthy Scottish nobleman called John Napier of Merchiston. Napier's computing system relied on his two greatest innovations: a set of calculating tables known as logarithms, and a calculating apparatus called Napier's Bones. Napier, who was educated at the University of St Andrews and later at various continental institutions, developed his inventions through his fascination with mathematics and, especially, theology.

In Napier's lifetime, religion had a grip on society that can be difficult to comprehend today. When Napier was born, Scotland was still a Catholic country. But during the subsequent Protestant Reformation and the accompanying civil war between forces loyal to the Catholic monarch Mary, Queen of Scots, and her Protestant son James VI (later James I of England), young Napier became a convinced – in fact, fanatical – Protestant. In his first major book, *A Plaine Discovery of the Whole Work of St John*, published in 1593, he applied logical and mathematical principles to solving a key problem troubling Protestant

intellectuals: how to calculate the date of the apocalypse, or end of the world. Napier scoured the *Book of Revelation* for information that would allow him to compute the answer. He finally calculated that the world would end in 1688 or 1700. In the process, Napier was also able to 'prove' that the Pope was the Antichrist.

From a modern perspective, it is easy to view this early work as a theological dead-end, even a misapplication of science for superstitious purposes, but Napier regarded it as his true vocation. The book was a bestseller throughout Europe. As if to try to make the last day come sooner, Napier wrote another book called *Secrete Inventionis*, or *Secret Inventions*, which contained details of various weapons of mass destruction. These included a giant mirror that could channel the sun's rays into a concentrated beam that would burn the Catholic ships of the Spanish Armada; a tank and a submarine; and a bomb that would annihilate everything within a wide radius. None of these inventions, thank God, ever got off the drawing board.

At the more peacable age of 64, in the house he had built for his family at Gartness on the banks of the River Endrick in Stirlingshire, Napier produced a truly great innovation: logarithms. Napier's logarithms were first presented and explained in his 1614 book, *Mirifici Logarithmorum Canonis Descriptio*, or *A Description of the Miracle of Logarithms*. The word logarithm comes from 'logos', meaning ratio, and 'arithmos', meaning number. Napier's written table of logarithms, which became known as Napier's Log, allowed the user to reduce complex calculations requiring multiplication and division of long numbers down to much simpler processes of addition and subtraction. In the process, Napier's logarithms popularised the use of the decimal point as a way of expressing complex fractions. In today's offices, electronic computers are able to carry out an infinite number of calculations without the need for a vast pile of written log tables, but the principle remains the same.

Napier's logarithms were an overnight sensation among the businessmen and scientists of the early modern world. Calculations that otherwise would have taken years to complete could now be done in a fraction of the time. Napier's log tables allowed seafaring merchants to calculate accurately their position on the oceans, just as they gave astronomers the means to calculate accurately the orbit of planets around the sun. The system of converting complex division and multiplication into much simpler arithmetic was applied to the scientist's other great invention: Napier's Bones.

Napier's Bones, or Napier's Rods, is a powerful pocket calculator. It has been refined over the years and in its standard

John Napier.

form consists of a case containing ten rods. Each rod has an individual number at the top from 0 to 9. The case also contains a board, numbered down the side 1 to 9, on which the rods can be arranged. Below the number at the top, each rod displays, in a vertical row down its spine, the double, triple, quadruple and so on, of that number. By varying the order in which the rods are put into the board – say 9124356780 – the answer to 'what is 5 x 9124356780' can be found by finding '5' on the side of the board, scanning horizontally across to pick up the quintuples of each number, and following some simple rules of carrying over, to produce the figure 45621783900. More advanced use of the grid allows for much more complicated calculations; dividing 9124356780 by 456231 to produce 19999.423, for example. Napier's Bones, which was made from metal, wood and sometimes actual bone, was a far more powerful apparatus than any abacus, largely because it was so much easier to use. Napier's innovation did owe a great debt, however, to the development of the abacus in other countries.

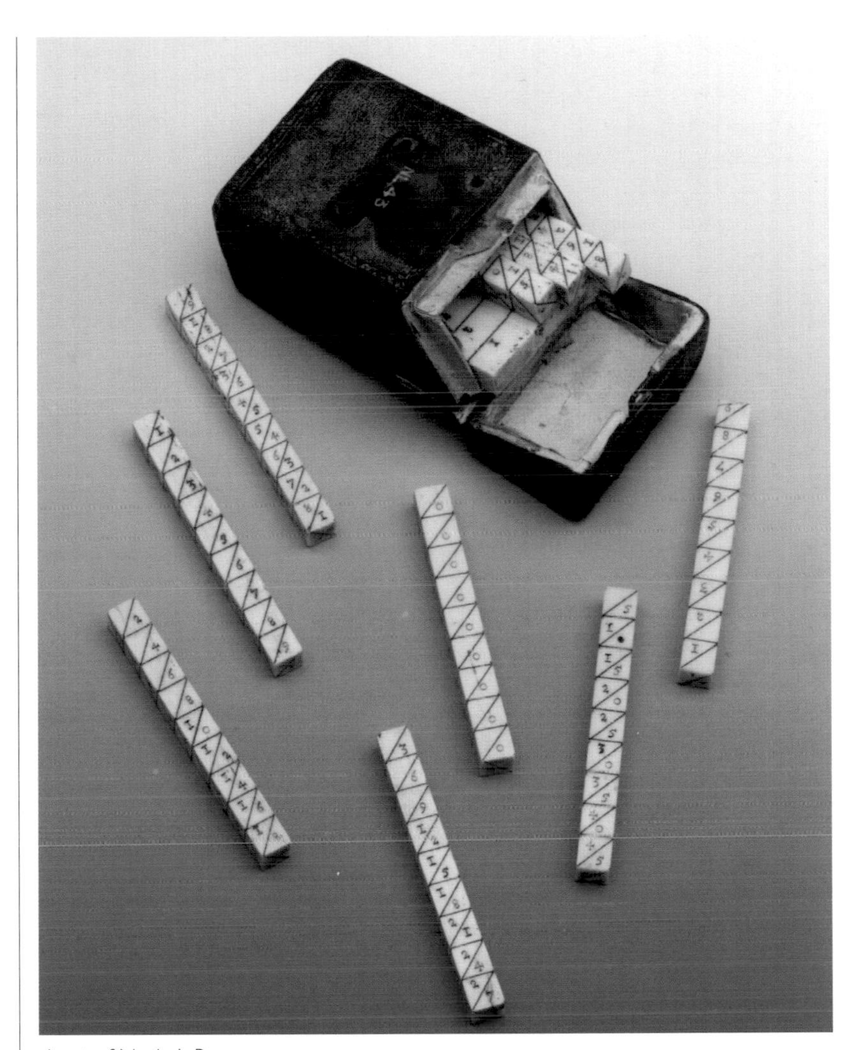
A set of Napier's Bones.

In his lifetime, many people regarded Napier's inventions as magical. Indeed he acquired a reputation as something of a sorcerer. When it became clear that one of his servants was stealing from him, Napier enlisted the help of a black rooster – one of two odd pets he kept, along with a spider in a box – to find the culprit. One at a time, each of Napier's servants was shut in a room alone with the bird. They had each been instructed to stroke the cockerel and told that it would crow when touched by the guilty party. The bird did not crow once, but Napier had secretly coated the bird's feathers in soot. He reckoned that the innocent servants would have no qualms about touching the bird, and so end up with dirty hands. But the thief would be too afraid to touch it, for fear of being caught, and would only pretend that he had. The thief was found to be the servant with the clean hands and was duly sent packing. The method was not exactly scientific, but it did demonstrate an originality of thought worthy of a great modern police detective.

After two marriages, and raising several children, Napier, who had devoted much of his time away from his study to improving his estates, died of gout in 1617 – perhaps brought on by cramming in too much earthly pleasure after inheriting his father's estates late on in life. He was buried in St Cuthbert's Church in Edinburgh. The city's Napier University is named after him, as is the Nepar crater on the Moon – using a spelling of the name used by Napier himself – and asteroid 7096 Napier.

Adam Smith, by John Kay.

ECONOMICS
Adam Smith, 1723–1790

If calculation is fundamental to business, then so too is the market. Of course, markets have been around since the Stone Age, when people would meet up to barter livestock and grain. But the modern market, which is more a set of principles than an actual place, and depends on the division of human labour into specialised industries and the free trade of goods and services in order to function, owes its origins to an eighteenth-century Scotsman from Kirkcaldy, Fife, called Adam Smith. According to the Adam Smith Institute, the powerful economic think-tank that now bears his name, 'Adam Smith was the pioneer of what today we call economics. He championed the benefits of specialization and free trade, creating the very idea of the modern market economy that dominates the free world today.' Not bad for an absent-minded professor who was known for his hypochondria and habit of talking to himself.

So Smith was the inventor of modern economics. But how do we quantify such an invention? What object or artefact can we hold up as tangible evidence? The answer is Smith's most important work, *An Inquiry into the Nature and Causes of the Wealth of Nations*, a multi-volume publication which took twenty-seven years to write and was finally completed in 1776, the year the United States of America declared its independence from Britain and set about becoming the world's leading economy. *The Wealth of Nations* was therefore composed at the dawn of the industrial revolution, at the beginning of the modern age, and used real-world examples to describe an economic structure that has remained essentially unchanged since.

While professor of philosophy at Glasgow University in the 1750s, Smith came into contact with the sons of wealthy colonial merchants who plied the Atlantic trade routes from the Clyde to Virginia and the West Indies. In this role, and later as tutor to the young Duke of Buccleuch, during which time Smith and the Duke spent several years on a Grand Tour of the continent, Smith observed how the spheres of economics and politics operated in the real world. Smith also helped the young James Watt set up in business, fascinated by the inventor's work with steam engines and other industrial innovations. During a two-and-a-half-year residence in Paris, Smith also became a disciple of the great French philosopher Voltaire and of the economist François Quesnay, who shared his interest in how the global economy worked.

Unlike many other intellectual works of the great Enlightenment era in which Smith lived, his *Wealth of Nations* is written in a clear prose style, demonstrating his talent for rhetoric, and remains easy to follow. Although he used neither the term 'capitalism', nor 'laissez-faire', it is today's free-market capitalist world that Smith's book foretells; a world whose dominant ideology argues that individuals acting in their own self-interest will ultimately create economic and social benefits for all, and that good governments should leave those individuals well alone to get on with their affairs instead of hindering them with interventions

and regulations. This is how Smith put it in what is, arguably, the most famous passage in *The Wealth of Nations*:

'Every individual . . . generally, indeed, neither intends to promote the public interest, nor knows how much he is promoting it. By preferring the support of domestic to that of foreign industry he intends only his own security; and by directing that industry in such a manner as its produce may be of the greatest value, he intends only his own gain, and he is in this, as in many other cases, led by an invisible hand to promote an end which was no part of his intention.'

Smith's description of an 'invisible hand' ensuring each self-interested individual unwittingly operates in a manner that is beneficial to his customers and to his fellow countrymen has been interpreted as evidence that the capitalist way of life, or the free market, is the religion of the modern age. The hidden hand belongs, if not to God, then to some higher power. However, the role of the 'invisible hand' in Smith's philosophy can be overstated. Elsewhere, Smith argued that what actually keeps us from harming society is our innate human nature, in which self-interest is counter-balanced by a sense of propriety that stops us from cheating or trying to steal from other people.

Smith himself was kept in line by his mother. He lived with her, unmarried, until she died at the grand old age of ninety. Among his close friends was the brilliant philosopher David Hume, with whom he enjoyed discussing a wide range of learned subjects while getting drunk and playing cards. Other than the story that, at the age of four, he had been briefly and mysteriously kidnapped by gypsies, little else is known about Smith's private life beyond the fact that he was known to be friendly, untidy and eccentric. If only he hadn't ordered his vast collection of private papers to be burned after his death we would know more about this brilliant and fascinating individual. He is buried in Canongate Kirkyard, Edinburgh, in a neo-classical tomb.

THE POSTAGE STAMP AND POSTMARK

James Chalmers, 1782–1853

A classic example of individual self-interest resulting in benefit for all is the story of the postage stamp and postmark. It is a common complaint today that the efficient running of offices and businesses of every kind is hindered by the unreliability of the postal service. This is as nothing, however, when compared with the situation that prevailed in the early nineteenth century. In those days, the delivery of mail was paid for not by the sender but by the recipient, under a complex and fragmented system of tariffs depending on how far the piece of mail was to be sent. The situation caused chaos, as people frequently refused to pay for mail that was sent to them, and the Post Office lost revenue as a result. It took a weaver's son from Arbroath, James Chalmers, to show that there was a better way.

Chalmers was an enterprising sort who left Arbroath and established himself as a printer, publisher and bookseller in Dundee, becoming a local politician and business leader in the process. In the 1820s he grew frustrated at how the slow and unreliable postal service of the time was hampering the growth of his own business. In order to do something about it, he first lobbied for improvements in the speed of deliveries between Dundee and London. The Post Office took note and the delivery time was shortened by a day.

But Chalmers had greater plans for reform. In 1834 he argued that a postage stamp should be introduced, by which a sender could pay for the mail, and which could be cancelled after it had been posted to prevent it being reused. He demonstrated his scheme by using his own printing press to print an adhesive postage stamp, with its value marked on the front, and attaching it to a letter which he posted to Rowland Hill, the secretary of the General Post Office. The stamp was crossed by a postmark to prevent it being reused. This is thought to be the first time in history that a letter carrying a stamp was posted.

Credit for this invention is often attributed to Hill, who in 1837 published a pamphlet *entitled Post Office Reform: Its Importance and Practicability*, which argued for the introduction of stamps. But it was Chalmers who made the idea of a stamp a reality. The innovation was adopted by parliament, and the Penny Black stamp, accompanied by a new set of post office regulations describing how the new system of prepaid stamps worked, was introduced in 1840. It revolutionised mail communication almost overnight. Today, an unused pristine Penny Black can fetch upwards of £1500 on the collectors' market. Unlike Hill, who was honoured and financially rewarded for his work on postal reform, Chalmers himself did not profit by his part in it, save for a plaque commemorating him as 'the originator of the adhesive postage stamp', which was erected by his son, Patrick, in 1888 on the site of Chalmers's Dundee bookshop.

STEREOTYPING
William Ged, 1699–1749

Another Scottish inventor who has perhaps not been given the credit he deserves is William Ged. In 1725, he invented the stereotype, a solid metal printing plate. For the first time, the stereotype allowed printers to reprint books quickly and easily without having to reset entire pages

from individual pieces of metal type. The plate was created by taking a plaster cast of the original, which was a matrix of individual type pieces. These were of the finest quality and imported from Holland. The cast was then used to produce a cheaper printing plate, also known as the stereoplate or stereotype. Ged also came up with a way of stereotyping delicate designs for the jewellery trade. In both these innovations, Ged was informed by his work with metal as a goldsmith in his home town of Edinburgh.

Ged was unable to find backing for his invention in Edinburgh, so he took it to London. There he entered into partnership with a stationer and a typesetter, and used his stereotype process to print an edition of the works of the classical Roman historian Sallust, and two prayer books. However, Ged's efforts to get the business off the ground were thwarted by the professional jealously of others in the printing trade, as well as the inability of his workmen properly to grasp the new technology. He returned to Edinburgh. Refusing offers from Holland to export his ideas there, apparently for patriotic reasons, Ged died in relative obscurity in October 1749.

The invention of stereotyping lived on, however. A number of innovators took up the idea later in the 1700s, including a French printer called Firmin Didot. Didot is thought to have been the first person to coin the term 'stereotype', though he was not, of course, the originator of the process itself. After that, stereotyping became a standard practice in printing because it allowed newspapers, in particular, to be run off at great speed and massively reduced cost – a process that has only relatively recently been superseded by electronic publishing. The word stereotype acquired a more widely-known meaning, of course, as a metaphor for any idea that is repeated identically and uncritically, especially when that repetition leads to ignorance of the complexities of the original. Similarly, the word cliché also comes from the stereotyping process. The French word meaning to stereotype, *clicher*, comes from the sound made when molten lead, the material they used to make the cast, came into contact with the matrix of metal type.

Stereotype plates.

THE ENCYCLOPAEDIA BRITANNICA
Macfarquhar, Bell, Smellie, 1768–1771

Next to a dictionary, the single most valuable work of reference on the shelf of the typical classroom is an encyclopaedia. If we accept that the more scholarly an encyclopaedia is, the better, then it can be fairly said that the world's pre-eminent encyclopaedia – certainly among English-speaking nations – was invented in Scotland. Published in three volumes between 1768 and 1771, this is the *Encyclopaedia Britannica*.

The first edition, which came into being in Edinburgh, was the work of 'A Society of Gentlemen'. Among the most important of those gentlemen were the printers Colin Macfarquhar and Andrew Bell, who came up with the concept and provided printing and distribution. Bell also contributed a number of illustrative copper engravings to the work, including scientific diagrams and artistic impressions, of which several were more lascivious than scholarly – an early example of the publishing maxim that sex sells. Perhaps the most important of all among the *Encyclopaedia*'s creators was the young writer and editor hired by Macfarquhar and Bell to provide the actual text, twenty-eight-year-old William Smellie.

Initially a printer to trade, Smellie was born in Edinburgh in 1740 and educated at the Royal High School. He was a gifted writer with an exceptionally wide range of interests – not to mention a taste for strong drink and the pleasures of the flesh, in which he is known to have indulged in later years with such friends as Robert Burns in the bars and brothels of Auld Reekie. Although Smellie's encyclopaedia prose became infamous for its opinionated style, numerous factual inaccuracies and several baseless theories, on the whole it was a quite brilliant summation of all the fields of human knowledge and endeavour – and helped make the *Encyclopaedia* a runaway commercial success.

Of course, the *Encyclopaedia Britannica* was an invention that drew heavily on earlier encyclopaediae. These included Pliny's classical Roman encyclopaedia, Ephraim Chambers's *Cyclopaedia* of 1728, and the *Encyclopedie* of Denis Diderot, published in Paris between 1751 and 1772. While some of these precursors were designed as dictionaries of arts and sciences, the work of Macfarquhar, Bell and Smellie was intended to be an even weightier work of reference. This aim is perhaps best illustrated by its full title: *Encyclopaedia Britannica, or, A Dictionary of Arts and Sciences, Compiled upon a New Plan, In Which the Different Sciences and Arts Are Digested into Distinct Treatises or Systems; And the Various Technical Terms Etc Are Explained as They Occur in the Order of the Alphabet.*

By the time of the third edition of 1801, by which time both Macfarquhar and Smellie had passed away, the *Encyclopaedia* had grown to twenty-one volumes. New editions continued to appear throughout the nineteenth century, with multiple contributions written by leading authorities in their fields, until the business was bought by American publisher Horace Everett Hooper. He moved production to the United States and published the 11th edition – regarded as a

Pages from the 1911 edition of *Encyclopaedia Britannica.*

landmark of scholarship and style – in 1911. While many other encyclopaedias have come and gone, the *Britannica* is still going strong, and continues to flourish in print and online. The digital *Britannica*.com benefits from the searchability and easy updateability afforded by an electronic format. In turn, *Britannica* has influenced other major online encyclopaedias, such as the user-generated *Wikipedia*.

As well as being a great invention in its own right, the *Encyclopaedia Britannica* sums up all the learning, the creativity, and excellence in education, that Scotland became famous for in the middle of the eighteenth century. It is one of the most important documents of what we now call the Scottish Enlightenment.

THE BLACKBOARD
James Pillans, 1778–1864

It is no overstatement that, without the invention of the blackboard, many of the other great inventions of the modern world might never have been devised. Take Albert Einstein's formulas, for example. Famous equations such as $E=MC^2$ were worked out and disseminated to students using a blackboard and chalks. And generations of little Einsteins across the globe have each benefited in their own small way from being able to quickly take in ideas, and have their own ideas presented to their peers, via the classroom blackboard.

Before the blackboard, teaching could be a hugely tedious task. There was no really effective means of communicating a written idea to the whole class all at once. Moreover, there was no way for the whole class to combine its intellect for the purpose of quickly solving written problems or brainstorming new concepts. Instead, teachers had to work their way around each student, writing down an equation or question on small, individual slate boards.

All that changed in the early 1800s. Step forward James Pillans, headmaster of the Royal High School, Infirmary Street, Edinburgh. While teaching his pupils geography at the High School, Pillans came up with the idea of the blackboard, and also introduced coloured chalks to complement the conventional white chalk, and thereby enable the blackboard to be used for more

sophisticated presentations.

Exactly how or when Pillans came up with the idea of the blackboard is not certain, but it is known that he grew fed up with the traditional method of copying work onto each student's slate and decided to take these individual slates and hang them on the wall side by side, like tiles. The result was one large, makeshift blackboard. Pillans then wrote geography notes across the slates in large letters so that all his pupils could read them at once. It was a brilliant example of one of Scottish education's most highly prized virtues – common sense.

With the basic idea devised, the next step was obvious. A single, really large piece of slate could be used to create a purpose-built blackboard, to be hung in schoolrooms everywhere. Thanks to a ready supply of slate across the British Isles – and Scotland alone had around two hundred slate quarries in those days – a classroom revolution began.

Pillans himself left the Royal High School in 1820 to become Professor of Humanity and Laws at the University of Edinburgh, a post he occupied until shortly before his death. He was remembered as a teacher with a profound belief in education, which he argued should be compulsory for all children.

In recent times, new technologies have emerged to rival Pillans's blackboard. The plastic whiteboard and felt-tip pens, the overhead projector and now the computerised whiteboard all have their place in the twenty-first-century classroom. And while all of these still follow the blackboard's underlying principle, the blackboard itself remains in use across the world as an invention that has stood the test of time.

THE HISTORICAL NOVEL
Walter Scott, 1771–1832

The Royal High School in Edinburgh can be credited with moulding not only the encyclopaedist William Smellie and blackboard creator James Pillans, but also the writer Sir Walter Scott. Scott is not exactly famous these days, but tourists assume he must have done something pretty impressive to be commemorated in stone by the towering, 200 foot neo-Gothic, rocket-like Scott Monument on Edinburgh's Princes Street.

That he did. For a start, Scott

has been credited – or perhaps discredited – with inventing a romantic image of Scotland that consists of tartan, kilts and clans. He did so principally through his writing. As important as Scott's contribution to Scottish culture is, however, his writing had an even greater, more global, significance. Scott was also the inventor of a literary genre now known as the historical novel. This is the genre that gave us Victor Hugo's *The Hunchback of Notre Dame*, Leo Tolstoy's *War and Peace* and Charles Dickens's *A Tale of Two Cities* – and it's still going strong. Well-known historical novelists writing today include Umberto Eco, Sebastian Faulks, Gore Vidal, Colleen McCullough and Hilary Mantel.

So what is the historical novel, anyway? And how did Scott come up with it?

Born in Edinburgh, Scott spent much of his childhood in the Borders region of Scotland, where, at his grandparents' farm, he became fascinated by a medieval structure nearby known as Smailholm Tower. The tower's dark, brooding character, and its sorrowful associations with wars such as the 1513 Battle of Flodden between Scotland and England, had a big impact on young Scott's imagination. He was taught to read by his aunt Jenny in the town of Kelso and was educated at the Royal High School in Edinburgh.

During this time Scott devoured history books, travel books and medieval chivalric romances such as *Beowulf*, Sir Thomas Malory's *Le Morte d'Arthur* and John Barbour's *The Bruce* – a poetic account of the life of Scotland's great warrior king, Robert the Bruce. All of this rich childhood and educational experience made the adult Scott – a lawyer by profession – ripe for producing something remarkable.

In 1814, Scott, by then a successful poet, completed his first work of prose fiction – the novel *Waverley*. It is the first historical novel. The book's fictional hero, Edward Waverley, is a young soldier with his head in the clouds. In 1745 he is brought north from England to stay with his uncle in Scotland. During the visit, Waverley gets embroiled in the Jacobite rebellion of Bonnie Prince Charlie and his Scottish Highland clans against the Lowland British, mainly English, government forces.

With its depiction of Edward Waverley as a man of feeling who is emotionally moved by his surroundings and experiences, *Waverley* contributed greatly to the modern romantic movement in literature. More importantly, though, *Waverley* demonstrates itself to be the archetypal historical novel. It does so in two key ways. First, it is a fictional adventure in prose set amid real-life historical events. Second, in

Sir Walter Scott.

much the same way as other writers might use satire or allegory, Scott's book uses history to comment on present-day, real-life concerns.

The conclusion of *Waverley*, which suggests England and Scotland should bury their historic differences and learn to get along with one another, while also demonstrating that peasants can be as noble as princes, is directed at the hot political and social topics – not of 1745 – but of Scott's own lifetime. In 1814, the relationship between Scotland and England was still tense and mistrustful despite a century of political Union – not least because the last Jacobite rebellion was only just passing out of living memory – while calls for greater respect and more rights for society's lower classes were now being heard.

Scott went on to apply the same technique to a string of hugely successful historical novels, including *Ivanhoe*, which did as much to arouse contemporary interest in England's relationship with its past and sense of identity as *Waverley* had done for Scotland's. Again, *Ivanhoe*, although set in twelfth-century England, pointed its finger at current concerns, with references to the ill-treatment of Jews. In 1819, when *Ivanhoe* was published, the question of whether religious freedom should be granted to Jews (and other minority religious groups, such as Catholics) was becoming a hot political issue in Britain.

Scott's literature was a huge international success, on a scale never before achieved by a living author. As we have seen, novels such as *Waverley* promped writers in many other countries to apply the same narrative principles to the histories, politics and national identities of their own countries. But Scott's other commercial ventures – including his interest in the business of publishing itself – fared disastrously by comparison.

Scott died in 1832 after working himself to death trying to pay his debts, but his legacy lives on. In the industrial Victorian era, *Waverley* lent its name to the main train station in Edinburgh, next to the site of the Scott Monument. It is a fitting tribute. The station is, after all, a place where the capitals of Scotland and England are connected, where all classes of people rub shoulders as equals, and where a thousand human dramas unfold every day.

SOCIOLOGY
Adam Ferguson, 1723–1816

Adam Ferguson, a Gaelic speaker and in later life a vegetarian, was a unique individual. A key mover, shaker and thinker of the Scottish Enlightenment, he was both sociable, hosting many parties and clubs and introducing the young Walter Scott to Robert Burns, and socially curious – to the extent that his invention is the science of studying how societies work, also known as sociology.

Ferguson was born in Logierait, Perthshire, and studied at the universities of St Andrews and Edinburgh. His degree in Divinity and his knowledge of Gaelic played a part in his enlisting in Highland regiment the Black Watch as an Army chaplain. In 1745 the young Ferguson was posted abroad with the regiment, where he took part in the Battle of Fontenoy in Belgium. It is said that he fought in the ranks and refused to leave the field of battle even when ordered to do so by his commanding officer. Ferguson stuck with the military life for a good few years, until eventually he got fed up with its paltry pay and conditions.

The experience of battle taught Ferguson some important lessons that he went on to apply in his next career as an academic. His intellect, sociability and gift for lecturing saw him rise through the ranks to become Professor of Natural Philosophy and then of Moral Philosophy, both at Edinburgh. Ferguson frequented the coffee houses and drinking clubs of Edinburgh, and sojourned often to the continent, becoming friends with such bright lights as David Hume and Voltaire. All the while he was developing ideas for his major work, *A History of Civil Society*, published in 1767.

This book was a great invention for a number of reasons. First, it saw Ferguson introduce the method of studying humankind in groups, or as different classes of people – an approach that formed the basis of the discipline we now call sociology. Second, also fundamental to sociology, was Ferguson's view that human beings are essentially social animals. Our behaviour is not shaped by some higher power, such as God or 'natural law', but by our interaction with other people. Third, and this is perhaps where Ferguson's front-line experience of battle comes in, the history of civil society, as he saw it, is essentially a story of conflict and strife between social groups – a proposition that later had a huge influence on the thinking of one Karl Marx.

Perhaps it was bloodlust, an attempt at living up to the principles of human conflict that he preached, but Ferguson later got himself embroiled in the American war of independence after being sent to Philadelphia to negotiate on Britain's behalf with the American colonists. He later returned to Scotland to settle down after suffering a stroke in 1780. He became a great advertisement for vegetarianism after switching to a meat-free diet and living to be almost 93 years of age. He died at St Andrews, leaving behind the aforementioned work plus his *History of the Roman Republic* of 1783 – regarded by many scholars as one of the most eloquent history books ever written.

James Hutton.

GEOLOGY
James Hutton, 1726–1797

Yet another alumnus of the Royal High School, James Hutton was the world's first rock star. We are, of course, talking about the science of geology, which was pioneered by Hutton in a number of printed works, including his *Theory of the Earth* of 1788. That work was 25 years in the making – a mere blink of the eye in comparison with the age of the Earth, which Hutton was the first to demonstrate conclusively as being much, much older than was previously thought. Or, as he put it, 'we find no vestige of a beginning, no prospect of an end.'

Hutton, who was born in Edinburgh and enjoyed stints in Paris and Leyden, spent much of his upbringing in Scotland studying rock formations on his family estate in Berwickshire. He later travelled the

Arthur's Seat, Edinburgh.

country looking for evidence to support his theories, finding significant rock formations, now known as Hutton's Unconformity, at Lochranza on the island of Arran.

As well as laying the foundations of geology, Hutton also wrote *The Theory of Rain*, which contributed much to the discipline of studying the weather, or meteorology. More significantly, Hutton's interest in the history of the Earth grew into an interest in the history and provenance of the creatures that populate it.

Upon the last subject, Hutton wrote: '. . . if an organised body is not in the situation and circumstances best adapted to its sustenance and propagation, then, in conceiving an indefinite variety among the individuals of that species, we must be assured, that, on the one hand, those which depart most from the best adapted constitution, will be the most liable to perish, while, on the other hand, those organised bodies, which most approach to the best constitution for the present circumstances, will be best adapted to continue, in preserving themselves and multiplying the individuals of their race.'

This was an evolutionary theory of natural selection, written more than a decade before Charles Darwin was born. The problem with Hutton, who died in 1797, was that he had an exceptionally ponderous and obscure writing style, turning out books whose 2000-plus pages were as impenetrable as the rocks he examined. Luckily, however, Hutton's writings were clarified and rewritten by later authors, such as the Scottish lawyer and geologist Charles Lyell, whose texts were studied by Darwin on board the *Beagle* as the vessel took him to the Galapagos Islands, where he found the evidence to support evolution.

TELESCOPE
James Gregory, 1638–1675

Alongside the blackboard, the rotating globe and the abacus – or perhaps Napier's Bones – the telescope surely ranks as one of the

A Gregorian telescope.

iconic classroom instruments. As suggested by its name, the Gregorian Telescope was invented by James Gregory, an astronomer from Aberdeenshire, in 1663. His design predated the famous telescope of Sir Isaac Newton, a friend of Gregory, and used a series of curved mirrors to reflect into the eyepiece the magnified image of an object a great distance away.

Although Gregory was unable to find a sufficiently skilled optician to manufacture an actual telescope during his lifetime, his design was so well thought-out that, not only did it successfully translate into a real working telescope in later years, but it was so good that the Gregorian telescope became the standard astronomical observing instrument for more than a century and a half. Even in recent times, Gregorian optics have been used in deep-space radio telescopes, and Gregory has been fittingly rewarded with having a crater on the Moon named after him.

William Ramsay.

NOBLE GASES

William Ramsay, 1852–1916

All of us who have attended science class surely remember Bunsen burners heating bubbling liquids to produce strange gases, or holding test tubes out of which rise wisps of coloured fog. And many of us must recall studying the Periodic Table of elements. But do we remember the answer to this question: what do argon, helium, neon, radon, krypton and xenon have in common? Answer: they are all 'noble gases', and they were first discovered by the Glaswegian scientist William Ramsay.

Ramsay's invention is his classification of the noble gases he had discovered, for which he won the Nobel Prize for Chemistry in 1904. Born in Glasgow and educated at Glasgow University, then Tübingen in Germany, Ramsay's talents saw him land a job as professor of chemistry at Bristol University and then University College London (UCL). It was at UCL that Ramsay – also a keen sportsman and music-lover – was inspired by his colleague Lord Rayleigh to search for an unknown element that could account for certain discrepancies among the known gases of the atmosphere.

Argon was the first of the new elements Ramsay discovered, and as he discovered more he gave them names that reflected their novelty – such as neon, which is classical Greek for 'new'; krypton, which means 'hidden'; and xenon, which means 'stranger' or 'alien'. After winning the Nobel Prize, Ramsay became a celebrity, and in 1905 endorsed a company that claimed it could extract gold from seawater. The scam did not much tarnish Ramsay's reputation, and in later years he did important work in the then pioneering science of radioactivity and nuclear physics. He died at High Wycombe, Buckinghamshire.

THE GRAND PIANO
John Broadwood, 1732–1812

For many, a highlight of our schooldays is the music class. And where would the music class be without a piano? The piano-maker John Broadwood, who was born at Oldhamstocks in East Lothian, gives his name to the oldest surviving piano company in the world, the prestigious John Broadwood & Sons. Trained as a carpenter, young Broadwood travelled – indeed it is said he walked much of the way – the 400 miles to London, where he found work in 1769 with the Swiss harpsichord maker Burkat Shudi.

Shudi was one of the greatest instrument makers around. In 1765, nine-year-old child prodigy Amadeus Mozart had been sent to London to play a Shudi harpsichord. But Shudi was more impressed by his young apprentice, Broadwood. So much so that he made Broadwood a partner in the business, and gave him his daughter's hand in marriage. That union produced four children and, after that first wife died tragically young, Broadwood's second marriage bore him six children, resulting in a sizeable brood to keep the piano business going – a business that Broadwood had taken control of after Shudi senior's death in 1773.

Broadwood set about improving piano design, shifting from the traditional box, or square, design to a new model, a prototype grand piano, which he perfected around 1777. Along the way, Broadwood patented the foot-pedal for modulating the sound, which did away with the clumsy levers that were then in use. He further improved the tonal quality of pianos by such innovations as the bass bridge.

Broadwood became so successful that he abandoned the manufacture of harpsichords altogether to focus solely on his pianos, which were sold throughout Europe. He extended the range of the instrument to produce the first six-octave grand piano in 1794. The music world was hugely impressed, and continued to favour this piano after Broadwood's death. Among the firm's clients have been some of the greatest piano composers and players ever, including Mozart, Haydn, Chopin, Beethoven and Liszt.

GOLF
The people of St Andrews, c.1100

Of all sports, golf is one of the oldest and most popular. It is played by many millions of people around the globe, on approximately 32,000 courses – with new courses being added every year. It is also a sport that, despite the best efforts of men down the years to keep the ladies out, has seen women rank among its most notable players: from Mary, Queen of Scots, who played golf in Musselburgh, East Lothian, in the sixteenth century, to Sweden's Annika Sörenstam, now one of the most successful golfers in history. The most successful golfer of the present day is, of course, an African-American by the name of Tiger Woods.

The huge popularity of golf among people of all nationalities, genders, races and creeds is something of which the Scots can be very proud – for it was, of course, in Scotland that the game was invented. As to exactly when the game was devised, this is less clear-cut. According to tradition, the first game of golf was played in the 1100s or thereabouts, on the sandy coastal soil of what is now the Royal and Ancient Golf Club of St Andrews. Shepherds used sticks to knock pebbles into rabbit holes and kept score. In time, on this and other early courses, the stones were replaced with balls, and holes were dug by hand to create a man-made course.

We know that golf must have been well established in Scotland by the 1400s, because the game was proving so popular it attracted the attention – or rather the irritation – of King James II. The first written account of golf being played was made in Edinburgh in 1457, when the king banned the playing of golf, since it was distracting the people from the more militarily useful practice of archery. That the ban had to be reiterated to the crown's disobedient subjects twice during the subsequent fifty years only goes to prove the game's extraordinary popularity.

Thanks to a busy maritime trade between Scotland and the Low Countries, medieval Scottish golf may have crossed the North Sea and influenced, or been influenced by, a game in the Netherlands called 'kolf'. The Dutch

game involved players striking a leather ball with a stick to hit targets in a field – although kolf targets were not holes, a crucial difference between that and true Scottish golf. In any case, these early games were not the finished article. Modern links golf as we know it was yet to emerge. When it did, it was a purely Scottish invention.

After the ban against playing golf in Scotland was lifted in 1502, it developed into the more advanced sport we recognise today. A variety of clubs were used to hit balls into a sequence of holes marked by flags on a specially designated, secluded seaside course, or 'links' as it is called in Scots.

As golf continued to grow in popularity, it found greater acceptance with the Royal Family. After the Protestant Reformation of the mid-sixteenth century, however, the game attracted the ire of the Church because the only day upon which people were generally free to play it was the sabbath. In the 1560s Mary Queen of Scots provocatively played golf, perhaps in order to irritate the Church elders who critised her 'Popish' love of games and frivolities, on the coast just outside Edinbugh at the Musselburgh Old Links Golf Course, which is today the oldest still-playing golf course in the world. In 1618, the curmudgeonly Church was slapped down by King James VI (and I of England), who decreed that the people were free to play golf on Sunday so long as it was not during times of service.

In 1744 the oldest surviving rules of golf were drawn up in Edinburgh. By the nineteenth century, golf had taken root in Scotland's neighbouring countries of England, Wales and Ireland, with new rules and terms introduced to what was quickly becoming an international sport – including formal restrictions on women players, which in some places still survive today. The game's popularity on the continent was hardly less pronounced; notably in Scandinavia, where Swedes were seen playing golf as early as 1830, and a strong golf-playing tradition quickly emerged. In Spain, La Puerta De Hierro golf club was founded in Madrid in 1895, two years after New Zealand staged its first amateur championships.

Of course, it is in the United States of America that golf truly found its second home. Fittingly, America's first great golf-course designer was a Scotsman called Donald Ross, who left his native Dornoch and emigrated to the States in 1899. Ross went on to design more than four hundred courses and, as a player, to achieve considerable success in the US Open. In keeping with Scottish tradition, Ross's course designs still followed the principle established at St Andrews several hundred years earlier. You play nine holes out from the clubhouse and then play them back again to reach the magic total of eighteen.

A boxing match fought under the Queensberry Rules.

BOXING

9th Marquess of Queensberry, 1844–1900 (with John Graham Chambers)

'No shoes or boots with springs allowed.' Just one of the twelve commandments in the sport of boxing, both professional and amateur, known as the Marquess of Queensberry Rules. Apart from banning excessively springy shoes, the rules also formalised the size of the boxing ring and other basic elements such as the wearing of gloves, the ten-and-you're-out count and the three-minute round. The rules are named after John Sholto Douglas, the controversial and eccentric

9th Marquess of Queensberry.

Born in 1844 in Florence, Queensberry was of noble Scottish ancestry and inherited the Queensberry estate in Dumfries-shire, southern Scotland, upon his father's death. He was a keen boxing and athletics enthusiast and, although a reactionary elitist in many respects, was also a champion of the rights of the common man to participate in sporting competitions. In 1866, Queensberry became one of the founding members of the Amateur Athletic Club, an association that was among the first to admit competitors from outwith the upper classes.

The following year, under Queensberry's patronage, the association published the Marquess of Queensberry Rules. Although the rules were principally devised by Welshman John Graham Chambers, a sporting polymath and superb athlete, without Queensberry's input and sponsorship it is unlikely that they would have been adopted as the standard code of practice for world boxing. Queensberry went on to further sporting achievements as a rider and racehorse owner, never losing his passion for boxing. After a long and colourful career as an outspoken atheist, politician and advocate of blood sports, he died in 1900 and was buried in Scotland.

BASKETBALL
James Naismith, 1861–1939

How do you keep a group of bored youths from tearing up the college campus in their free time when the winter is too harsh to set foot outside? Answer: you invent basketball. This was the task accomplished in 1891 by James Naismith, a physical education teacher at the Springfield Young Men's Christian Association in Massachusetts. Naismith, who was born in Ontario, Canada, to Scottish parents, was instructed by the head of his department to ensure the game did not take up too much room and was not too rowdy or rough. To that end, Naismith worked out that the game needed a large, soft ball – he used a football initially – and a set of rules that involved minimal physical contact.

The key innovation that made the new game a safe indoor sport was the creation of a keeperless goal, or basket, high above the players' heads, which Naismith devised by hanging a couple of peach baskets from the gallery ten feet above the floor of the college's gym hall. In later years, purpose-built baskets were introduced, along with a specially designed basketball. Naismith also drafted the thirteen rules of basketball, which, although revised and updated, still form the basis of the modern game. The National Basketball Association hall of fame, which is also located in Springfield, is called the Naismith Memorial Hall of Fame in his honour.

HIGHLAND GAMES
Traditional

The Highland Games are as iconically Scottish as kilts, bagpipes and whisky – and fittingly all three of those things features heavily at any decent gathering. But what really draws the crowds are the games' unique sporting events. While there is a great deal of athleticism required of Highland dancers who compete at the modern games, central among the traditional sporting elements are the shot put, or stone put, the caber toss and the hammer throw. These three events are regarded as Scottish sporting inventions, in that their origins can be traced to the emergence of the first Highland Games, when the clans-men of Highland Scotland were called to a gathering by their chief in order to

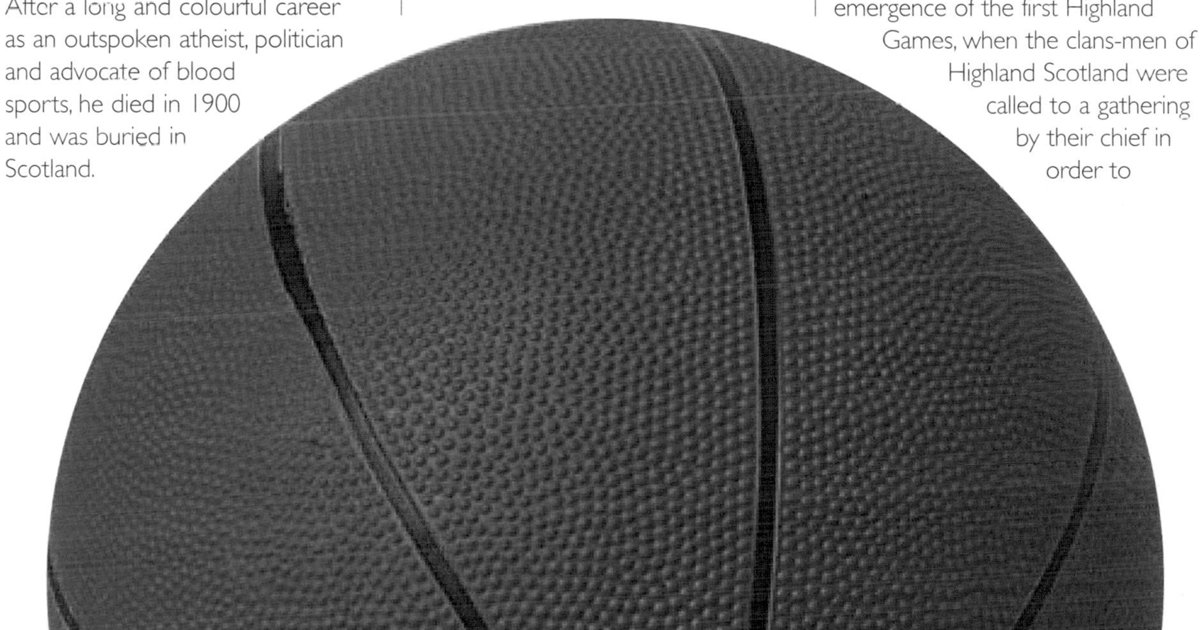

practice their military skills and hone their physical prowess – and in the process provide a pleasing spectacle for the rest of the clan.

Exactly when the games and their associated sports came into being is a matter of debate. It is often said that the Highland Games as we know them today, along with the small kilt and modern Highland dress, are an invention of the Georgian and Victorian eras. Indeed, the games grew hugely in popularity after the Braemar Highland Games was patronised by Queen Victoria from 1848 onwards. In many respects the games do owe a lot to nineteenth-century innovations, especially with regard to the formalising of clan gatherings with office-bearers, written rules, dress codes, calendars of events, fees, and so on. But if we think in terms of a more ad hoc Highland gathering, then folkloric evidence strongly suggests that these events have a long history stretching back to medieval times – the era of traditional clan society.

Whatever their precise origins, the caber toss, hammer throw and stone put are rightly regarded as traditional Scottish sports. In the caber toss, competitors must carry a full-length log, usually made of Scots pine, balanced vertically in the interlocked palms of their hands and cradled by their neck and shoulder. The aim is to run forward and then toss the caber so that the upper end swings forward in an arc and hits the ground, with the lower end then following up and over until the caber is tossed through almost three hundred and sixty degrees. Successful competitors are said to have 'turned the caber', and the competitor who comes closest to keeping the forward trajectory of the log in the twelve o'clock position is judged to be the winner.

The hammer throw features a hammer made from a metal ball weighing around 20lbs, attached to a wooden pole or handle. The contestants use the handle to whirl the hammer around their heads and then launch it as far as they can. In the stone put, or shot put, the contestants throw a stone of about 20-26lbs in weight as far as they can. The stone is delivered from a position standing on the spot, or after a short run-up, depending on the rules of the competition. In both cases, hammer throw and stone put, the contestant who throws the furthest wins.

These Highland sports have made their mark far beyond Scotland, with Highland gatherings held in countries as far and wide as Canada and New Zealand. And it is generally accepted that the modern Olympic shot put event owes something to the traditional Highland stone put. It remains to be seen, however, whether the tossing of the caber will ever become an Olympic sport.

Above left. Throwing the hammer.
Above right. Highland dancing.

AT THE BANK

CENTRAL BANK
William Paterson, 1658–1719

Scottish banking at the end of the first decade of the twenty-first century found itself in a sorry state, as once noble institutions were struck down by a poisonous cocktail of unregulated greed, arrogance and stupidity. It is perhaps fitting, therefore, that the long and colourful history of Scottish banking began at the end of the seventeenth century in similar circumstances – amid a climate of hubris, avarice and financial blunders of Atlantic proportions. Looking back, what is truly remarkable is that the Scots dug themselves out of a financial mess three centuries ago with a variety of ingenious fiscal inventions, which, in the process, effectively laid the foundations of modern banking.

The great Scottish banking inventions begin, somewhat ironically, with the Bank of England, the world's first central bank. It was devised by William Paterson, an enterprising farmer's son from Skipmyre near Lockerbie. As a young man, Paterson emigrated first to England and then to the Bahamas to make his fortune. In 1694, Paterson, by then a wealthy trader in the West Indies, argued that the English government needed a central bank, an institution that could control and regulate England's rapidly growing imperial economy and repair the then dire state of the kingdom's public finances. Paterson outlined his proposal in his pamphlet, *A Brief Account of the Intended Bank of England*. It was persuasive: the Bank of England's royal charter was granted in July that year.

Paterson proposed that the bank get off the ground by issuing a loan of £1.2 million to the English government. In return the bank's subscribers were designated the

The Bank of England in 1790.

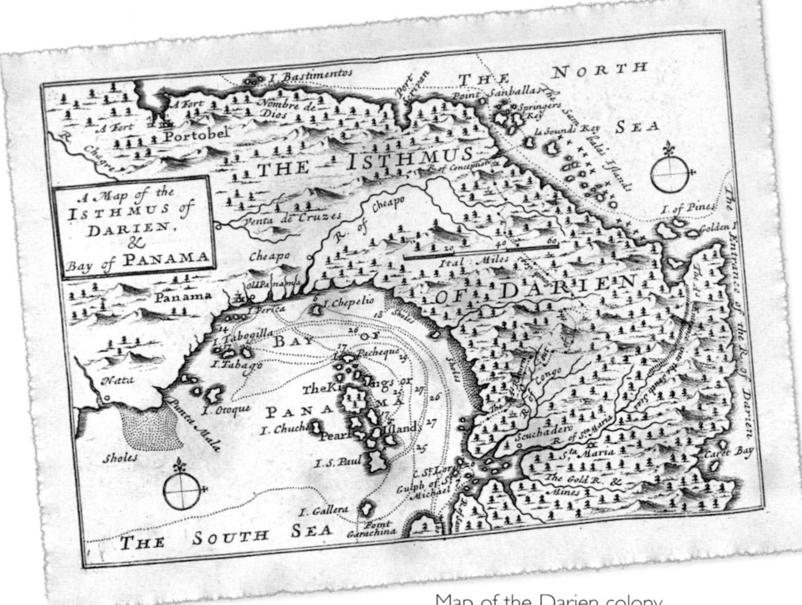

Map of the Darien colony.

Governor and Company of the Bank of England. Over time, the bank evolved and its functions came to include controlling interest rates, maintaining currency stability, protecting the British economy from financial threats, managing the country's gold reserves, assisting the government of the day in the implementation of its financial policies, and, as demonstrated recently, being a lender of last resort to other economically vital institutions, such as large private banks, at risk of collapse. The template was copied the world over, and today the US Federal Reserve and the European Central Bank are among the most powerful of central banks.

Flushed with his success at setting up the Bank of England, Paterson went on to be instrumental in the establishment of the central bank of his native Scotland – then still an independent kingdom – in 1695. The Bank of Scotland was followed by Paterson's next venture, the infamous Darien scheme. Using his knowledge of central America, Paterson proposed a trading *entrepot* on the strategic Isthmus of Panama, the setting up of which required £400,000 of investment. This huge sum was to be raised solely from within Scotland, because investors in London refused to back the scheme, fearing competition with England's East India Company. The ships set sail in 1698, but the venture was doomed. Had circumstances been different, Paterson had chosen the perfect location to launch Scotland as an independent imperial power. But due to lack of sufficient funding, coupled to other major setbacks, the ambitious scheme failed, with Paterson losing his wife and son to the ravages of tropical disease in the colony. Scotland was almost bankrupted in the process.

Scotland's economic collapse at Darien paved the way for the incorporating Union of 1707 with England, as the Scots joined forces, albeit very grudgingly, with their commercially superior neighbour. The United Kingdom of Great Britain went on to become the world's strongest economy in the two centuries that followed, thanks in no small part to the steadying influence of Paterson, who died in 1719, and the Bank of England.

THE OVERDRAFT
Royal Bank of Scotland, eighteenth century

The overdraft was invented in 1728 when the Royal Bank of Scotland, then just a year old, allowed merchant William Hog to take out £1000 more than he had in his account. The Royal Bank was set up because the principal private bank then operating in the country, the Bank of Scotland, was thought to be sympathetic to the Jacobite cause which threatened the Hanoverian monarchy and Union with England.

The Royal Bank was established as the bank of Hanoverian loyalists, its first governor being Archibald Campbell, Lord Islay, a politician and keen gardener whose portrait is reprinted today on the front of all Royal Bank of Scotland banknotes.

Right. An ATM.

Below. Scottish banknotes.

THE CASH MACHINE AND PIN NUMBER
John Shepherd-Barron, 1925–2010

The Automated Teller Machine (ATM), or cash dispenser, was devised in a 'Eureka!' moment of which Archimedes himself would have been proud. The ATM's inventor, John Shepherd-Barron, thought it up while sitting in the bath. That was back in the 1960s, and the first machine was installed in 1967 at a London branch of Barclays bank. There are now more than one-and-a-half million ATMs installed around the world – a world in which it is sometimes difficult to imagine how people managed before the magical 'hole in the wall' appeared on our high streets.

A Scot born in India, Shepherd-Barron graduated from Edinburgh and Cambridge universities before going on to work in London at De La Rue, the banknote and paper manufacturer. He rose to become managing director of De La Rue Instruments and, during that time, came up with the ATM. Although a mechanical dispenser had previously been developed in America in the late 1930s, it was not effective and was withdrawn from service after only a few months of operation at the City Bank of New York. So Shepherd-Barron's machine, which predated plastic cards and instead 'read' a radioactive code from a special cheque that the customer inserted into the machine, was the first true and successful cash dispenser.

Shepherd-Barron's design required the customer to enter a personal identification number (PIN), which the the machine then matched to the code on the cheque, in order to receive what was then a fairly sizeable maximum of £10. Barron initially intended the PIN code to comprise six digits, but abandoned this because the longest sequence of numbers his wife could remember was four. Thus, thanks to Shepherd-Barron's wife, Caroline, each of us carries an easy to recall four-digit PIN code around in our heads today.

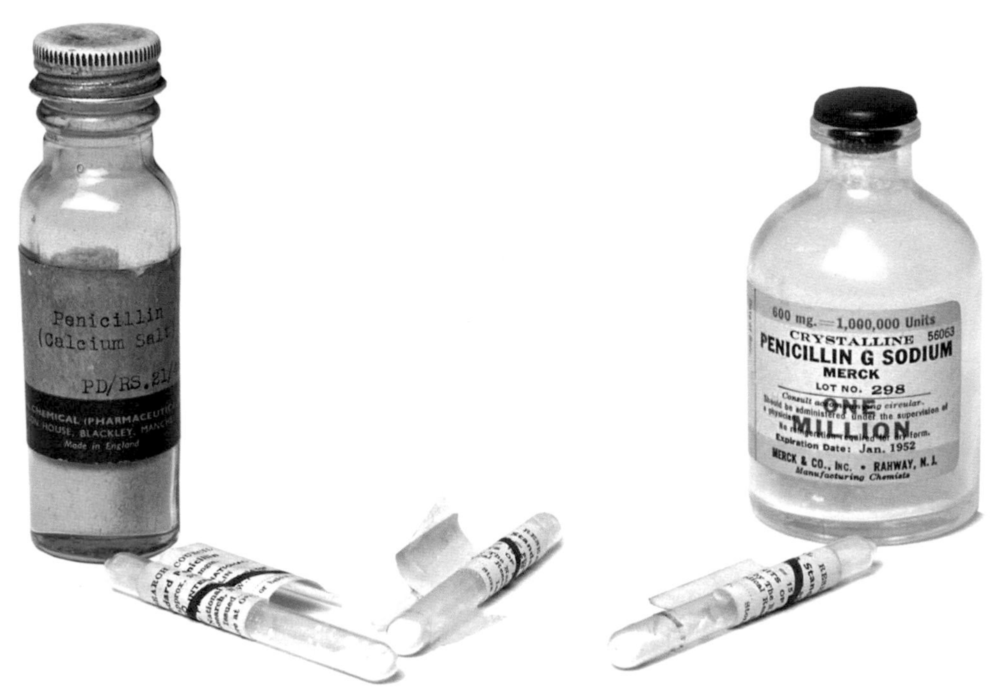

VACCINES
Various inventors, c.1700–c.1950

VACCINE AGAINST SCURVY
James Lind, 1716–1794

Ever wondered how the British got the nickname 'limeys'? It's a fascinating story, and it begins with a Scottish doctor and seaman by the name of James Lind. He developed a vaccine against scurvy, a disease which causes severe bleeding of the gums, tooth loss, hallucinations and, if left untreated, death.

Lind began his medical career in 1731, when he took up an apprenticeship in his native Edinburgh at the Incorporation of Surgeons – today known as the Royal College of Surgeons. He went to sea in 1739, becoming a ship's surgeon's mate during an era of colonial expansion in which the sea was one of the main battlefields between the emerging great powers. Lind saw action in the English Channel, the Mediterranean, the West Indies and Guinea. But to work in the navy was also, in those days, to work on the frontier of medicine. While many a bloody naval battle was fought between Great Britain and the other imperial powers of the time, it was observed by Lind and others that poor hygiene, malnutrition and disease killed many thousands more sailors than did combat or shipwreck.

Lind observed that scurvy was among the very worst of maritime diseases, capable of wiping out an entire ship's crew, and he became determined to find an antidote. He knew that before any attempt could be made to find a cure for scurvy, he first had to find out its cause. It became clear to him that the poor and extremely limited diet eaten by sailors on long voyages was in some way to blame, since scurvy was also found in others who were forced to eat a monotonous diet for many weeks and months, such as people living in castles under siege.

Exactly what the problem was with the sailors' diet became the focus of Lind's attention while he was serving on board the Royal Navy vessel HMS *Salisbury* in 1747. He conducted an experiment in which a dozen sailors, all suffering from scurvy, were selected from the ship's crew. The men were divided into six pairs and each pair was fed the same basic diet every day, but with some key differences.

To the diet of one couple, Lind added a quart (just over a litre) of cider. Another pair were given half a pint (almost a third of a litre) of seawater. Two other men were given a few spoonfuls of vinegar. A further two were given elixir of vitriol, which is an alternative name for sulphuric acid. One pair were treated to a paste made from garlic, mustard and horseradish, washed down with barley water. And finally, two sailors were instructed to eat a daily portion of citrus fruit, each consisting of two oranges and one lemon.

Result? Within a week, one of the men on oranges and lemons had made an almost complete recovery and was fit for duty, with the other not far behind. The two on cider, which was of course made from fruit, showed signs of improvement. The rest showed no change in their condition. Although the concept of vitamins was not yet established in Lind's day, the conclusion was clear: give the men a portion of citrus fruit daily – or fruit juice boiled down into a handy concentrate – and the navy need never worry again about losing sailors to scurvy. Lind retired the following year and later, in 1753, published *A Treatise on Scurvy*, which documented his findings, before being appointed chief physician at the Royal Naval Hospital in Portsmouth.

Incredibly, it was not until 1795, a year after Lind's death in Gosport, near Portsmouth, that the supply of lemon juice became a requirement of all vessels in the fleet. The slow uptake of his findings – which were supported by other physicians – is an indication of how little the Royal Navy valued the lives of its sailors in those days. Nevertheless, thanks to Lind, who had also advocated various improvements to general hygiene and wellbeing on board ship, the eventual introduction of lemon juice into the seafaring diet saw scurvy almost completely disappear. When lime juice began to be used as a cheaper alternative it gave rise to an enduring nickname – Limeys – for British sailors.

Lind left the world with three invaluable medical innovations. While the health benefits of citrus fruit had been vaguely known about for centuries, Lind was the first to clearly identify it as a cure and vaccine against scurvy. Secondly, his experiment on the sailors of HMS *Salisbury* is widely regarded as the world's first clinical trial of a medicine. Finally, Lind's general commitment to *The Most Effectual Means of Preserving the Health of Seamen in the Royal Navy*, as a 1757 essay of his put it, contributed to a huge improvement in the way authorities such as the army, navy and the government treated the health of the men and women under their wings.

VACCINE AGAINST MALARIA
George Cleghorn, c.1716–1789

Malaria is probably the greatest plague in human history. People have been attempting to come to terms with the disease's deadly effects since the first recorded references to it in China almost five thousand years ago. Yet, annually, malaria continues to claim the lives of around two million people – the majority being infants in Sub-Saharan Africa. But there is hope. A number of highly effective treatments have been developed in

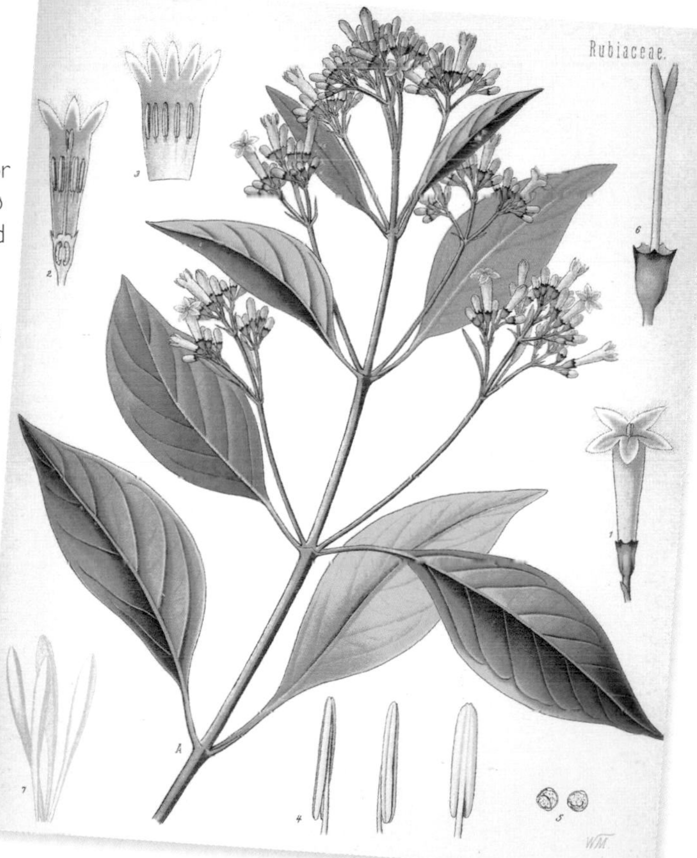

The cinchona tree, source of quinine.

recent years. And while these drugs are often unavailable to many of those in greatest need, it is hoped that there will soon be a return to the example set by the first great anti-malaria medicine – which proved effective, relatively cheap and readily obtainable. That medicine was quinine. It is made from the bark of the cinchona tree, and its use as an anti-malaria drug was devised by George Cleghorn, a farmer's son born in Granton, on the north side of Edinburgh.

After attending his local parish school, Cleghorn studied medicine at the University of Edinburgh before enlisting as an army surgeon in Minorca, in the Mediterranean Balearic Islands. It was there that Cleghorn discovered quinine's efficacy in treating malaria – a disease later found to be caused by parasites carried by Anopheles mosquitoes – and wrote about it in his book *The Diseases of Minorca*. While Cleghorn retired from the army to practice medicine in Dublin, his advocacy of quinine helped establish it as the principal weapon of attack in the war on malaria – a term that derives from the medieval Italian for 'bad air', 'mala aria'.

Quinine, as both a preventative medicine and a cure, helped eradicate malaria from Europe by killing what later became known as the Plasmodium Falciparum malaria parasite – although quinine is not without its problems, being extremely toxic if taken in too large or too sustained a dose. A range of more sophisticated medicines are now used in the fight against malaria, but quinine remains a powerful remedy when other alternatives are not available, thanks to the work of Cleghorn.

TREATMENT FOR TUBERCULOSIS
Sir John Crofton, 1912–2009

In folklore, tuberculosis sufferers were thought of as vampires. Their red, swollen eyes, sensitivity to light, pale skin and coughing up of blood made people believe that the victims needed to suck the blood of others to replace what they had lost. And in Victorian times, there was even a certain romance to tuberculosis, or TB. It was believed that the illness somehow made artists and writers more creative.

The truth about TB is far from romantic. Caused by a killer strain of airborne bacteria, it has claimed the lives of millions of people around the world. But many more would have suffered an untimely death from TB were it not for the work of such scientists as Sir John Crofton. Although Crofton was born in Dublin, it was during his long tenure as Professor of Respiratory Diseases and Tuberculosis at the University of Edinburgh, in the 1950s, that he devised a supremely effective treatment for TB.

Crofton's innovation was a therapy that comprised the simultaneous use of three different antibiotics. It quickly resulted in a huge decline in TB cases. He rose to become President of the Royal College of Physicians in Edinburgh and, in 1977, was awarded a knighthood to honour his life-saving treatment for this deadly disease.

ANAESTHESIA
James Young Simpson, 1811–1870

Once upon a time, the barber didn't just cut your hair – he cut your arms and legs off too. In medieval times, amputation was often the only way of giving a person a chance of recovering once a limb was seriously diseased. Of course, having a part of

The after-effect of Simpson's chloroform experiment.

your body chopped off is a serious enough business. But just imagine you had to stay awake during the operation.

That's how it was in right up until the nineteenth century, before the invention of anaesthesia. Even though barbers had given way to more modern surgeons, operations were still primitive and pain-relief almost non-existent. If an infection picked up during the procedure didn't kill you, then there was a good chance that the shock of being fully conscious while your limb was sawn off already had.

It was all too much for one young physician by the name of James Young Simpson. The son of a baker from Bathgate, West Lothian, Simpson studied medicine at Edinburgh University – but almost dropped out because he couldn't cope with the butchery of the Victorian operating theatre. It wasn't unusual for hysterical patients to be held down by five men while the surgeon sawed into their flesh as they lay fully conscious, writhing in agony. Not only was it a terrifying ordeal for the patient, it made the surgeon's work far more difficult than if the patient was still.

But Simpson stuck with his studies. After graduating, he went on to put himself well within screaming distance of many female patients by specialising in midwifery, or childbirth. He advanced quickly and, at the age of 29, was made Professor of Midwifery at Edinburgh. Prompted by his experiences of surgery and the agony endured by women during childbirth, Simpson decided to try to come up with a drug that would block or 'anaesthetise' – patients' pain.

During his search for an anaesthetic, Simpson heard about a drug called ether, which was being used in the United States. But when Simpson experimented with it, he found it had a nasty smell and it irritated people's mouths, noses and throats. He then came across chloroform. This drug was used in factories as a solvent to separate different substances, or to stick other substances together. But it was also known to make factory workers drowsy if they sniffed it by mistake.

Before trying out chloroform on any patients, Simpson again experimented with it himself. In November 1847, he conducted a test at his home. He and two of his assistants sniffed a bottle of chloroform and became unconscious. To the relief of the others present, the men later came round, feeling fine, and Simpson had a 'Eureka!' moment. He realised that chloroform anaesthetic would be a safe and effective way of ending centuries of unnecessary suffering on the operating table – and during childbirth.

A few days later, Simpson delivered Wilhelmina Carstairs, a doctor's daughter. In so doing he achieved two world firsts: he was the first person to use anaesthesia in childbirth, and the first person to use chloroform in medical practice. Around the same time, a young boy on the operating table was administered chloroform by Simpson before having a chunk of diseased bone cut from his arm. In both cases, the mother and the boy slept soundly through the procedure.

Word of the benefits of Simpson's work spread quickly. One woman even named her child Anaesthesia in his honour, and Queen Victoria gave birth to her eighth child, Prince Leopold, under chloroform anaesthetic. The royal birth was an important vindication of Simpson's innovation, because it quickly silenced opposition to anaesthesia from churchmen. They had argued that anaesthesia was against God's will, because God had decreed that Eve and all women who came after her would experience pain during childbirth. Over time, however, there did emerge some scientific opposition to chloroform, when it became clear that excessive exposure could be harmful or even fatal.

Nevertheless, in Simpson's day, the benefits of chloroform were found vastly to outweigh the risks. It was used as an anaesthetic in more than a hundred thousand operations during the Crimean War of the 1850s and the American Civil War in the 1860s. For his efforts, Simpson was awarded with a baronetcy by Queen Victoria, an honorary degree by Oxford University and the freedom of the city of Edinburgh.

PENICILLIN
Alexander Fleming, 1881–1955

The town of Darvel in Ayrshire is small, but with some big claims to fame. Early industrialists made the place famous for manufacturing lace and muslin cloth. Long before that, the hero of Scotland's war of independence, William Wallace, fought a battle here. And long before that, the ancient Scots erected a monolith here known as the Dagon Stone, said to have life-giving powers. But it is the life-giving powers of one particular Darvel resident, Alexander Fleming, that are perhaps of greatest significance.

Penicillin in a petri dish.

Fleming, who grew up on Lochfield farm just outside the town, came up with something that many people regard as the greatest invention since the wheel: penicillin. It is hard to overestimate how many lives have been saved by this, the world's first antibiotic, ever since Fleming discovered it, named it and proposed its use as a medicine in the 1920s and 30s. Penicillin has proved an extremely effective treatment for blood poisoning, pneumonia, diphtheria, scarlet fever, typhoid and many other contagious diseases.

This remarkable achievement was the consummation of Fleming's career, which he spent in search of an effective anti-bacterial medicine. Perhaps surprisingly, medical practice was not the young Fleming's first choice of profession. He originally left his native Scotland to work in London in a shipping office, but abandoned that after just a few years to follow the example of his elder brother, Tom, a doctor. Enrolling at St Mary's Hospital in Paddington, Fleming quickly demonstrated his gifts as a physician, graduating with distinction and later winning several awards and medals. He was taken on as staff by St Mary's as soon as his studies were completed.

In 1914, with the outbreak of war in Europe, Fleming enlisted as a medical officer. It was in field hospitals in France that Fleming first became greatly interested in, and upset by, the germs that got into soldiers' wounds and caused their deaths – even after the surgery itself had been a success. Years later, in 1928, while working in his laboratory at St Mary's, Fleming happened to examine a Petri dish that he had thrown in the bin. He had previously been using it to grow blood-poisoning bacteria for examination. Now he noticed a mould growing in the dish. But he also noticed something much more remarkable. Around the mould was a clear area with no germs, as though the bacteria had been killed by a substance, a kind of juice, leaking from the mould. After further examination, Fleming named this substance 'penicillin'.

Although Fleming had discovered penicillin by accident, his achievement should not be underestimated. Even though he was later very modest and self-deprecating about his discovery, without his expert knowledge and understanding of what he had in front of him, the value of penicillin as a medicine would never have been realised. Penicillin would have remained nothing more than a mouldy liquid, destined for the rubbish tip. Fleming presented medical papers about his discovery, and its potential applications in medicine, but he lacked the necessary skills in chemistry to isolate penicillin and turn it into an actual medicine. That was left to two other researchers, spurred on by yet another world war.

Alexander Fleming.

In 1941, Howard Florey and Ernest Chain, two biochemists at Oxford University, succeeded in isolating penicillin and turning it into a practical drug. Their efforts were much admired by Fleming, who was knighted in 1944, and the three shared the Nobel Prize for Medicine in 1945. After Fleming's death, his laboratory at St Mary's became home to the Alexander Fleming Museum.

HYPODERMIC SYRINGE
Alexander Wood, 1817–1884

Ranking alongside the stethoscope and thermometer, the hypodermic syringe is an icon of modern medicine. Many of us can remember getting our first jags, or jabs, as schoolchildren, to protect us against various diseases, or to take samples of our blood. And surely we have all seen how hospital TV dramas from *Quincy ME* to *ER* have found the syringe a handy prop to heighten the tension, as a cocktail of drugs is injected into a patient's bloodstream in a desperate bid to save their lives.

Amazingly, for all that the hypodermic syringe is a very modern innovation, its origins go back more than a thousand years. In ninth-century Egypt, a hollow needle made from a glass tube was used to suck cataracts out of patients' eyes. That is the earliest known 'syringe', which comes from the ancient Greek word for tube. But it was not until the nineteenth century, during Victorian times, that a syringe was created with a pump action and needle point refined enough to pierce the skin and inject substances safely into a patient's bloodstream. This is where the term 'hypodermic' comes from, which means 'under the skin'.

The hypodermic syringe was invented by an Edinburgh physician called Alexander Wood. In the early 1850s, at his practice in the city's prestigious New Town, he began experimenting with a hollow needle and pump as a means of administering morphine as a painkiller for patients suffering from neuralgia, or nerve disease. He injected his first patient in 1853 and wrote about its success a couple of years later in *The Edinburgh Medical and Surgical Journal*. In later years, Wood's invention was taken up by others and used widely in intravenous injections (injections into veins), which is one of the fastest ways to get a medicine working on the body.

Wood's invention had some tragic unforeseen consequences. The first recorded fatality from a drug overdose administered by hypodermic syringe was Wood's wife, who overdosed on morphine. Yet without Wood's work, countless others would have died because medicines could not be administered to them effectively enough, making his syringe a great life-saving invention.

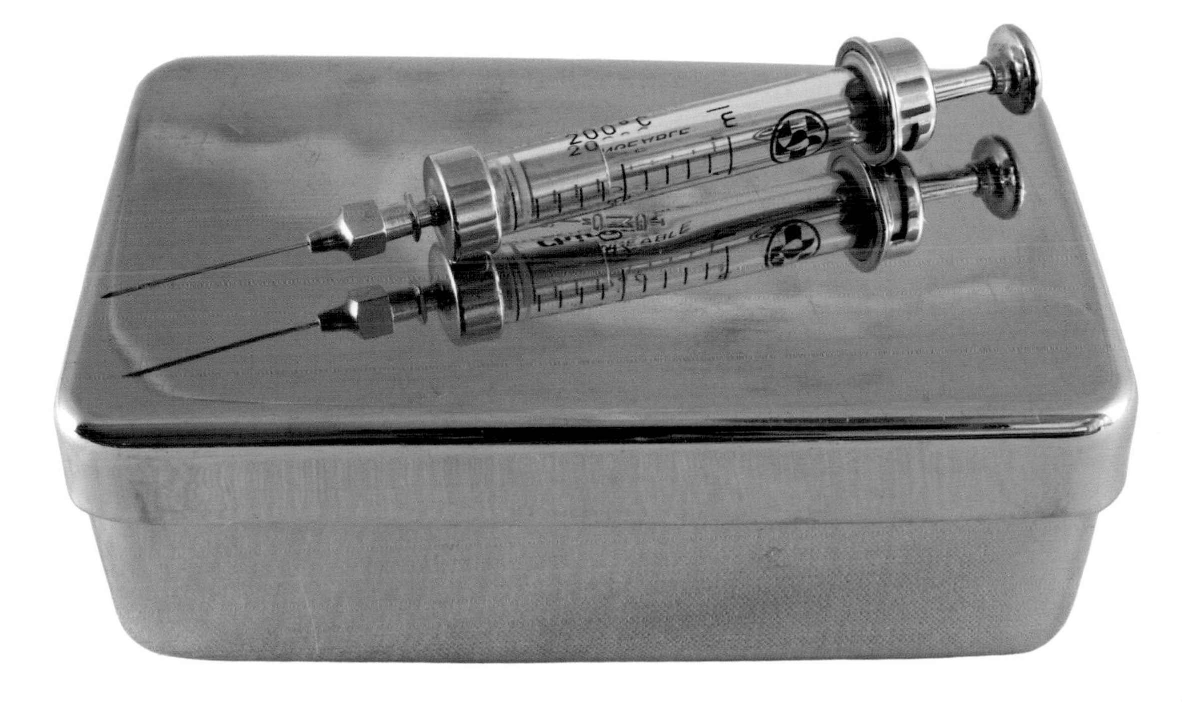

HYPNOTHERAPY
James Braid, 1795–1860

Keep your eyes on the bright, shining object . . . you are feeling very sleepy . . . very sleepy . . . Yes, we are all familiar with the idea of hypnotism from TV, films and literature. But did you know that its use as a medical treatment was invented by a Scot called James Braid?

Long before Braid appeared on the medical scene in the 1800s, hypnotism had been practised for centuries. It was believed by many people that the universe contained a supernatural magnetic fluid, or force, which influences the health of the human body. Hypnotists were supposedly able to make us of this force through a process called animal magnetism, which allowed them to put people into a waking sleep, similar to somnambulism, or sleep-walking, in which the subject would respond to the hypnotist's powers of suggestion. There was also some credible evidence that hypnotism could help cure illnesses.

Nevertheless, by the nineteenth century, many modern scientists and doctors were deeply sceptical about hypnotism. They regarded it as nothing more than superstitious nonsense. Yet some medical scientists realised that the apparently powerful effects of hypnotism could not be ignored, and that science had a duty to try to explain what was really going on. The most important of these curious physicians was James Braid, inventor of the medical discipline that became known as hypnotherapy.

Born at Ryelaw in Kinross, Braid was the son of a well-heeled landowner. The young Braid moved to Edinburgh's port of Leith to become a surgeon's apprentice, and studied medicine at Edinburgh University, gaining his diploma in 1815.

By that time he had married, and later had a son and a daughter. After working as a surgeon in a Lanarkshire mining community, Braid set up a private practice in Dumfries in 1825 before moving again, this time to Manchester, where he was to live for the rest of his life, and where he would devise the branch of medical science that made his name – hypnotherapy.

Braid became interested in hypnotherapy in 1841, when he watched a show by the French-Swiss hypnotist Charles Lafontaine. Inspired by Lafontaine, Braid conducted his own hypnotism experiments and came to the conclusion that hypnotism could be used as an effective medical therapy. He gave a series of lectures to demonstrate his ideas and later wrote them up in a book called *Neurypnology, Or, The Rationale of Nervous Sleep*.

Braid's innovation was his appliance of rational, scientific principles to something that had hitherto been steeped in magic and superstition. Whereas practitioners such as Lafontaine believed hypnosis relied on animal magnetism, or the transfer of 'mesmeric fluid' from the hypnotist to the patient, Braid demonstrated that hypnosis induced changes that were all within the patient's body and could be explained scientifically.

Braid showed that hypnotism put his patients' nerves 'to sleep'. In so doing, it caused changes in their blood flow and circulation, which had beneficial effects. For example, he found that hypnosis made the patient's muscles rigid. But when he applied pressure to a particular limb, say an arm, it ceased to be rigid and blood flowed into it with renewed vigour, improving bodily circulation. Braid also believed that, under hypnosis, the patient's mind, stimulated by appropriate suggestions from the hypnotist, could have a direct influence on particular bodily functions and organs – a process of 'mind over matter'.

So how did Braid hypnotise his patients? His approach was remarkably similar to the clichéd image familiar to many of us. He found that a hypnotic state was most easily brought on by keeping the patient fixated on a small bright object held above their eyes and about 12 inches away from their face. Patients could be released from hypnosis by blowing on, or lightly pressing, their eyeballs. Braid made use of this technique in his own medical practice.

As his understanding of hypnotherapy improved, Braid revised his ideas, writing articles and pamphlets such as *The Power of the Mind over the Body* (1846), *Magic, Witchcraft, Animal Magnetism, Hypnotism, and Electro-Biology* (1852), and *Hypnotic Therapeutics, Illustrated by Cases* (1853). After his death, Braid's work influenced later hypnotherapists, such as a group of French doctors who became known as the Braidists for their hypnotherapy experiments in the 1860s.

The Braidists were forerunners of a group called the Salpêtrière school in Paris. It was through that group, in the 1870s, that the science of hypnotherapy became properly established. But they acknowledged that the foundations of their work had been laid by Braid.

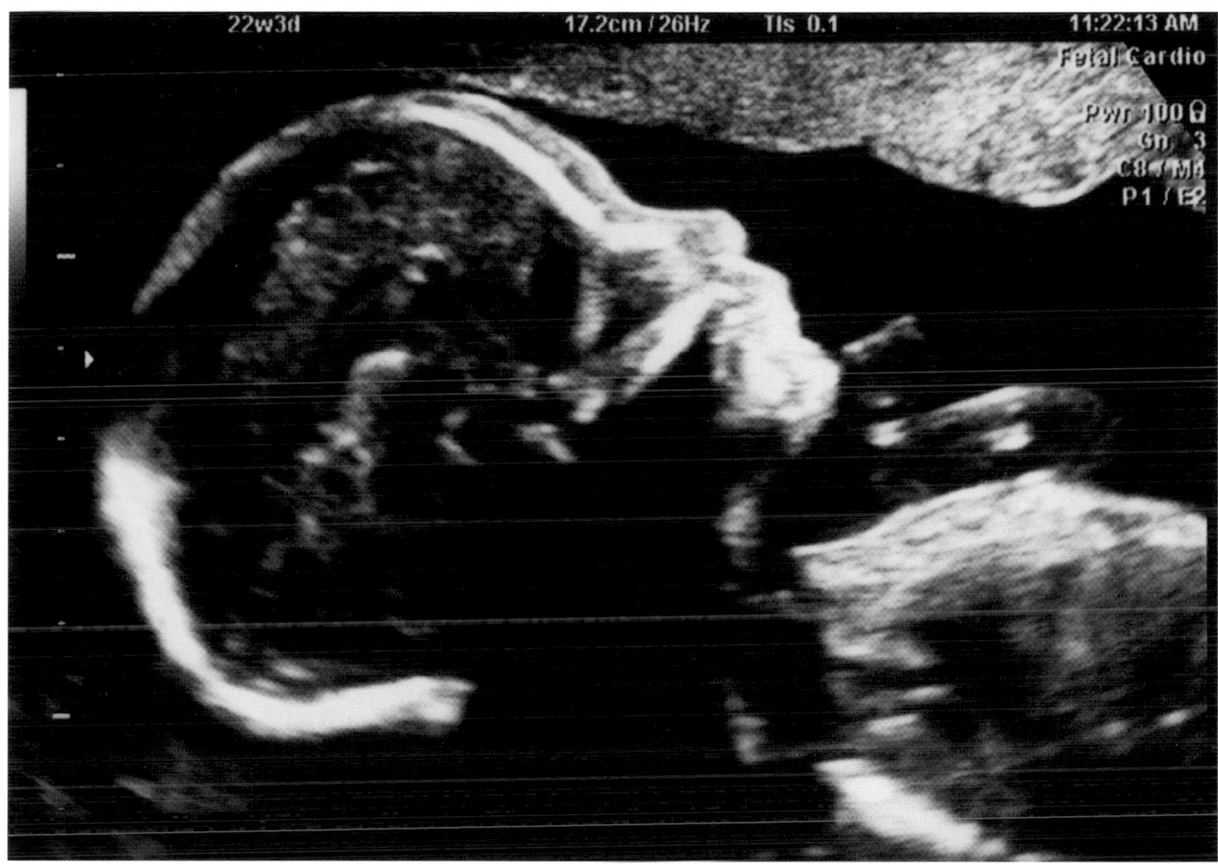

An ultrasound scan.

ULTRASOUND SCANNER
Ian Donald 1910–1987

What do bats, fish and pregnant mothers have in common? They all use ultrasound to 'see' what's going on. While the first two have a natural ability to pick up on ultrasound waves, allowing them to navigate in darkness or deep under water, when it comes to humans, we have to rely on an invention that has played a very special role in the lives of many people who were born in the past few decades. Thanks to the ultrasound scanner, invented in the late 1950s by Ian Donald while professor of midwifery at Glasgow University, parents today enjoy the privilege and peace of mind of seeing their unborn infants developing in the womb.

Donald, who was born in Cornwall but grew up in Scotland, served with the RAF during the Second World War and was recognised for his bravery. He qualified as a doctor at St Thomas's Hospital, London, where, in 1952, he gave a taste of what was to come by inventing a respirator for resuscitating new-born babies. A few years later, having taken up his professorship at Glasgow, his mind turned back to emerging technologies that had left a lasting impression on him during the war – radar and sonar.

Could the ultrasound waves used by sonar navigation systems in ships and submarines be harnessed for medical use? Donald believed they could, and borrowed a device from Clydeside engineering firm Babcock & Wilcox to test his theory. The device used ultrasound, or ultrasonic, waves to detect flaws in pieces of metal. But Donald believed that it could be run over a person's skin to reveal abnormalities in the human body, such as tumours. He conducted experiments using organic specimens with the machine and found that his hunch had been correct. The viewing screen on the device clearly showed flaws and differences in the tissues.

Donald believed a medical

ultrasound scanner would not only be of potentially great help in the treatment of various illnesses, such as cancer, but also in the diagnosis of problems in pregnancy and childbirth. With the help of Glasgow electronics company Kelvin Hughes, Donald produced a prototype diagnostic ultrasound machine in 1957. The following year, he reported on the machine's success in leading medical journal *The Lancet*.

Initially, there was some ridicule at the idea that a metal detector could be adapted to scan unborn foetuses. But the machine quickly won over a sceptical medical establishment. Donald was presented with a string of awards and honours for his work, and by the time of his death in the late 1980s his ultrasound scanner was a permanent fixture in maternity wards around the globe.

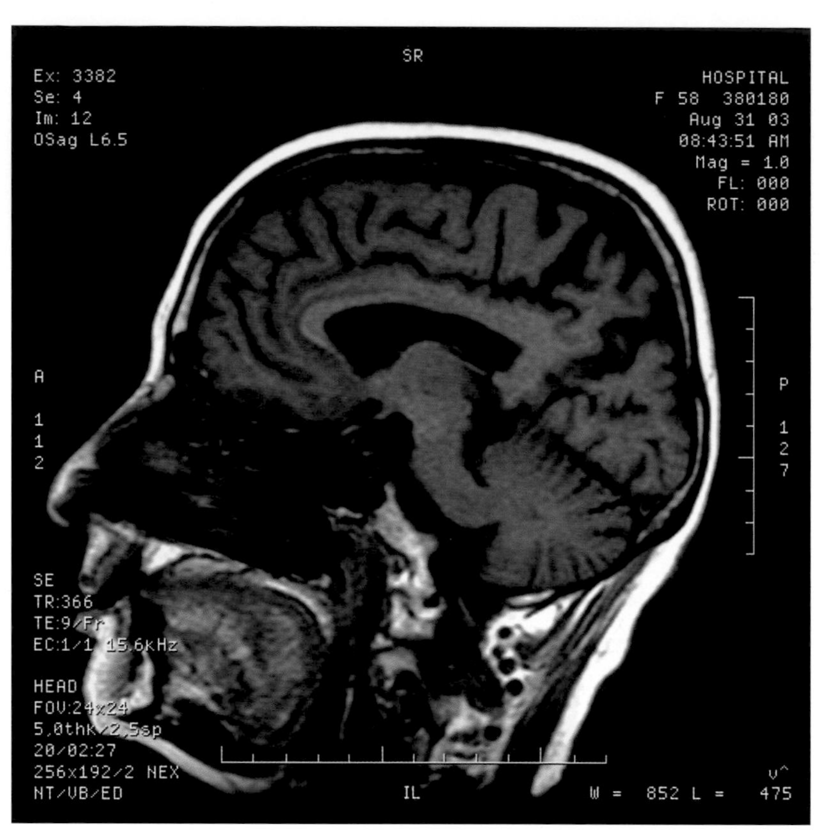

An MRI scan.

MRI BODY SCANNER
John Mallard, 1927–

If the medical ultrasound scanner was a landmark, then no less significant is the Magnetic Resonance Imaging (MRI) scanner. Although the design has changed since the first whole-body MRI scan on a patient was conducted at Aberdeen University in 1980, the classic MRI scanner looks like something from a science-fiction film.

Inside the MRI scanner's huge, doughnut-like tube, the patient's body is subjected to a hugely powerful magnetic field. That field manipulates the hydrogen nuclei, or protons, found in water molecules in the patient's body. Different tissues in the body react to the field in different ways, and those differences are transmitted to the scanner, which uses them to build up an image of what the inside of a person's body looks like. The MRI scanner is especially useful for examining the brain and internal organs, and helps in the early diagnosis and treament of cancer and heart disease.

The Aberdeen team was led by John Mallard, who moved there from London's Hammersmith Hospital in 1965 to become Scotland's first professor of medical physics. When he arrived in the northeast, Mallard had already built a primitive nuclear scanner which his colleagues used to detect brain tumours. After the team went on to build the first true whole-body MRI machine, trials in the early 1980s saw around a thousand Aberdeen patients scanned and images produced that made Mallard's team known throughout the world, and orders for MRI machines were placed by other hospitals.

The MRI scanner has not been without its probems. Although Mallard set up a company to market his invention, the huge costs of development and manufacture of MRI machines stymied progress – as did safety fears around nuclear radiation, the effects on implants such as pacemakers, and the fact that the hugely powerful magnetic fields could turn fairly innocuous metal objects in their vicinity into potentially lethal missiles. Meanwhile, just as the safety of the machines and their usage was being improved, rival versions of the technology began emerging from the United States, which has served to create the false impression that MRI technology was solely invented in America.

Whatever its precise origin, the MRI scanner is now an invaluable multi-billion-pound global industry. Coupled with huge ongoing advances in computer technology, it has grown into an impeccably safe, highly respected and life-saving technology that is undergoing continuous improvement, with colour images of extraordinary precision and clarity now being produced, and the portability of the machines much improved. And for all that, Mallard and his team deserve the principal credit.

DIALYSIS
Thomas Graham, 1805–1869

We've all used gel, right? Hair gel, toothpaste gel, soap gel . . . you know the sort of thing. Gluey, gloopy stuff that's not quite liquid and not quite solid. But did you know that the word 'gel' was coined by a Glasgow-born chemist called Thomas Graham? He used it to describe a type of colloid.

A colloid is a mixture in which one substance is dispersed in another. Gel is only one example. A colloid could also be part solid and part gas, like smoke. Or it could be a liquid mixed with another liquid, like the butterfat and water that makes up milk. A colloid is different from a solution. In a solution, the particles of the mixed substances dissolve together, whereas in a colloid the particles remain separate.

Graham, who became professor of chemistry at the University of London, used the principle that colloids are different from solutions in his great invention – a process known as dialysis. It is used today in hospital kidney dialysis machines.

A dialysis machine works by removing waste (which is soluble) from the patient's blood (which is a colloid). This separation is achieved using a membrane, or artificial skin. The membrane is porous; in other words, it contains tiny holes. Only the soluble waste can pass through these tiny holes, and as a result it seeps out of the blood, so the blood is purified.

Of course, Graham did not have the technology to build a dialysis machine. He used a more primitive apparatus called a 'dialyzer'. Nevertheless, it was an effective enough invention to demonstrate the basic principle.

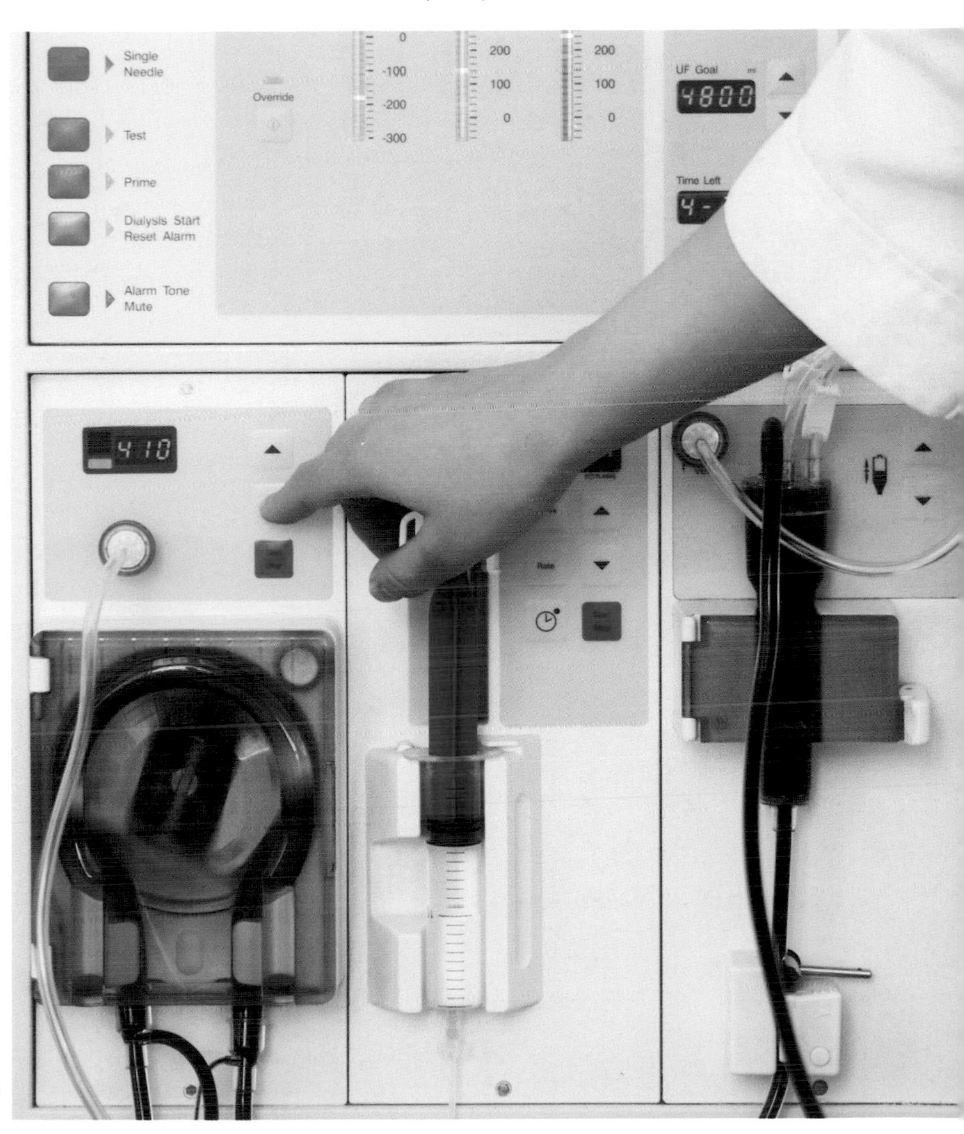

A dialysis machine.

THE BETA-BLOCKER
James W. Black, 1924–2010

With a name that sounds like a futuristic weapon, the beta-blocker has proved highly effective at combating illness. The first beta-blocker, a drug called propranolol, was unveiled in 1964, and since that time it has helped save the lives of millions of people, while improving the quality of life for millions more. The medicine works by 'blocking' unwanted or harmful activity in the body, particularly in the heart, greatly improving the conditions of patients suffering from angina, hypertension or a heart attack.

Propranolol, which became the template for many other beta-blocker drugs, was developed by James Black, the son of a coal-mine manager from Fife. Black attended high school in Cowdenbeath before going on to St Andrews University to study medicine. After a spell in Malaya, Black returned to Scotland in 1950 and set up the Physiology Department at Glasgow University. It was there that he developed his beta-blocking invention.

Black went on to a long and illustrious medical and academic career at several prestigious universities. Later inventions by Black included cimetidine, another massive medicinal step forward, which is used in such drugs as Zantac to treat heartburn and stomach ulcers. For his work, Black was knighted, made chancellor of Dundee University and, in 1988, awarded the Nobel Prize for Medicine.

THE NUCLEUS
Robert Brown, 1773–1858

How can an invention be both tiny and huge at the same time? If it is the invention of a concept known as the 'nucleus' of cells, that's how. It was devised by Robert Brown, the son of a churchman from Montrose, near Aberdeen. Brown was a keen botanist and, while conducting observations and experiments with plants, he established that the cells of every living thing contained a tiny body at their centre.

Brown made his discovery in 1831, while he was Keeper of the Botanical Department of the British Museum in London, and devised the term 'nucleus' for the little body he observed. Nucleus is Latin for 'nut' or 'kernel'. But that wasn't all. Brown was also the first person to produce evidence for the existence of atoms – the tiniest bits of matter in existence, from which all other things are made – when he used his microscope to observe pollen particles moving around at random in water. Although Brown did not realise it at the time, the movement was caused by atoms colliding with the pollen particles, causing the particles to jitter. This phenomenon, named in his honour, is known as Brownian Motion.

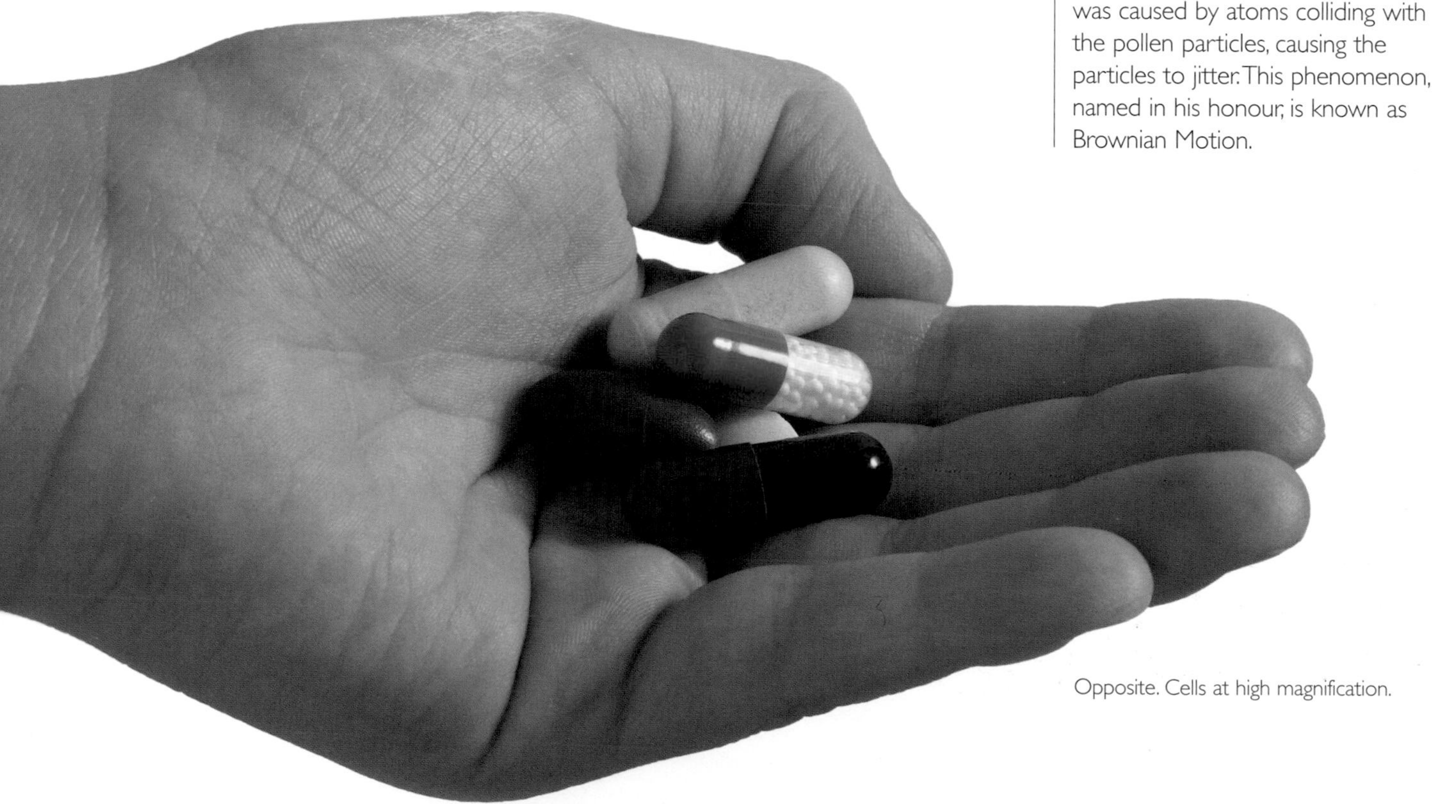

Opposite. Cells at high magnification.

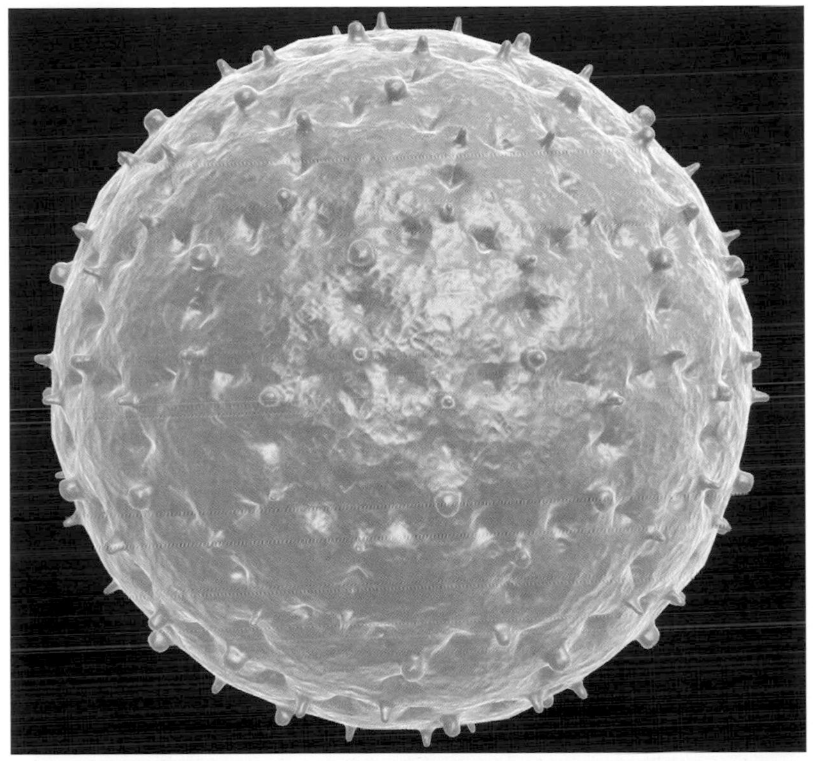

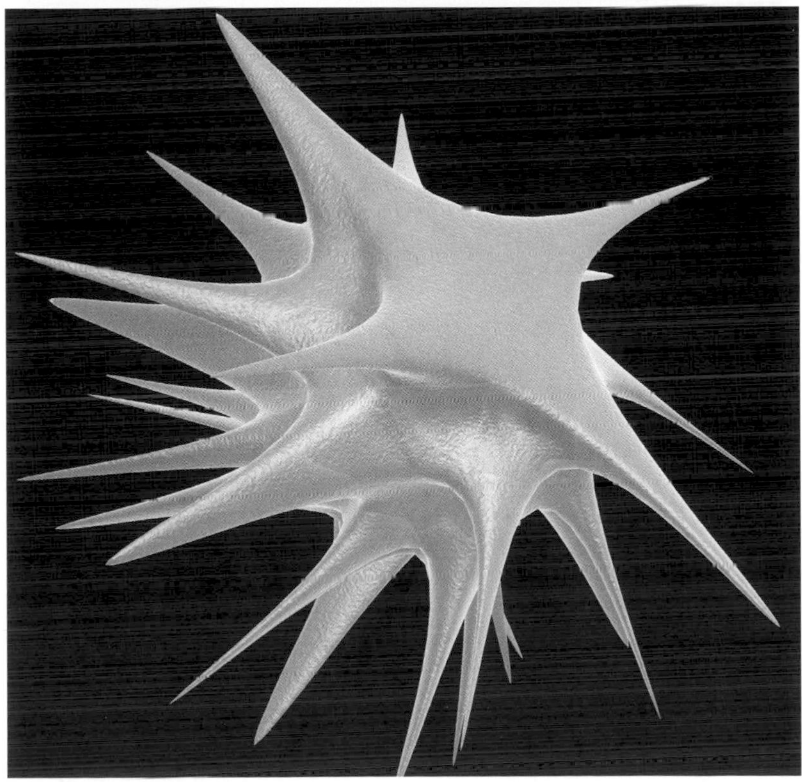

CLONING
Roslin Institute, 1996

The discovery of the nucleus of cells was the first step on a journey that has brought humankind to the brink of an entirely new stage of evolution. Just as many people at the dawn of modern medicine hoped, or feared, would happen, human scientists have learned how to 'play God'. In other words, we now know how to artificially create life.

Yes, human clones are here. Or at least, they could be quite soon, thanks to the Roslin Institute. The Roslin Institute, a part of the University of Edinburgh located on the southern fringe of the city, specialises in research into animal biology. Scientists at the Institute invented the technique that successfully cloned the first fully grown mammal in July 1996.

Cloning, which has long been the stuff of science-fiction dreams or Frankenstein nightmares, depending on people's points of view, is the act of making an exact, living copy of another person, animal or plant. Of all the medical advances of the past century, there is probably none more amazing or controversial than this. Trust the Scots to do it again.

The creation of the first cloned animal, a ewe named Dolly the Sheep, whose existence the Roslin Institute announced publicly in February 1997, was not the work of any one person alone. Most of the credit has been apportioned by colleagues to Professor Keith Campbell, an English-born biologist who spent much of his childhood in Perth, before obtaining a degree in microbiology at London University and PhD at Sussex University. He joined the Roslin institute in 1991

Dolly the Sheep.

and, along with his colleagues, began working towards the creation of a cloned animal.

So how did they do it? Dolly was created using a technique known as 'nuclear transfer'. In the Roslin laboratory, the nucleus of a cell taken from the udder of a six-year-old Finn Dorset ewe was implanted in an unfertilised egg, which had had its own nucleus removed. The result was an embryo that could grow into an exact clone of the donor ewe.

A couple of dozen such embryos were then implanted into thirteen surrogate ewes. Such a large number was necessary because the experimental nature of the technique made it likely that the vast majority of the embryos would not survive. Luckily, a hundred and forty-eight days later, one of those ewes, a Scottish Blackface, gave birth to Dolly.

In 2003, aged six, Dolly died, having given birth to five healthy lambs of her own – conceived the 'natural' way. There is a debate about whether Dolly died because her cloning led to premature aging, since ewes of her type can expect to live for eleven or twelve years. Some scientists believe Dolly simply succumbed early to a common lung disease, which was brought on by her having to spend too much time indoors for security reasons.

Since Dolly, scientists at the Roslin Institute have made many other important breakthroughs in this field. Elsewhere, horses, bulls, dogs and other animals have since been successfully cloned. But what of human clones? In 2008, researchers in America announced that they had successfully created the first viable cloned human embryos using cells taken from the skin of a donor adult human. The group added that the embryos were later destroyed before they had the chance to develop into foetuses.

Where this technology will lead in the future is anyone's guess. At the very least, it is likely to be used in some way to create cloned organs and other tissues to replace diseased ones, thereby saving or even prolonging the lives of millions of people.

Diseases such as cancer and Parkinson's could become a thing of the past. Cloning might one day result in human beings being able to live for ever . . .

THE STEAMSHIP

Patrick Miller, 1731–1815 and William Symington, 1764–1831

On 14 October 1788, the world's first steamship made its maiden voyage by chugging across the tranquil surface of Dalswinton Loch in Dumfries and Galloway. At a speed of five miles per hour, this extraordinary, paddle-wheeled vessel left all other forms of transport in its wake. It started a revolution in travel that would eventually lead to the Victorian golden age of ocean steamers, and then on to today's cruise liners. The Dalswinton craft was a joint invention by William Symington, a hard-drinking Lanarkshire engineer, and Patrick Miller of Dalswinton, a gentleman who claimed to have introduced the turnip to Scotland.

The two men got together out of shared interest. Miller was a wealthy banker and former seafarer with a penchant for dabbling in inventions. Among his many projects was a double-hulled pleasure boat, which was powered by paddle wheels mounted in between the hulls. The wheels had to be cranked by hand, which made travel very slow and laborious. So Miller hit upon the idea of using steam propulsion instead. And that's where Symington came in. He was working on a steam-powered carriage that could transport people across land, but when Miller approached him to design the world's first paddle steamer he jumped at the chance and dropped the steam-carriage project.

Miller was introduced to Symington, the son of a mining engineer, through a mutual acquaintance. Symington constructed a new ship by modifying Miller's paddle-ship design to accommodate a small steam engine that he had manufactured in an Edinburgh brass foundry. The result was a vessel twenty-five feet, or seven-and-a-half metres, in length, with a metal-plated hull. When its paddle wheels were successfully turned by the engine, propelling the craft forward, most of

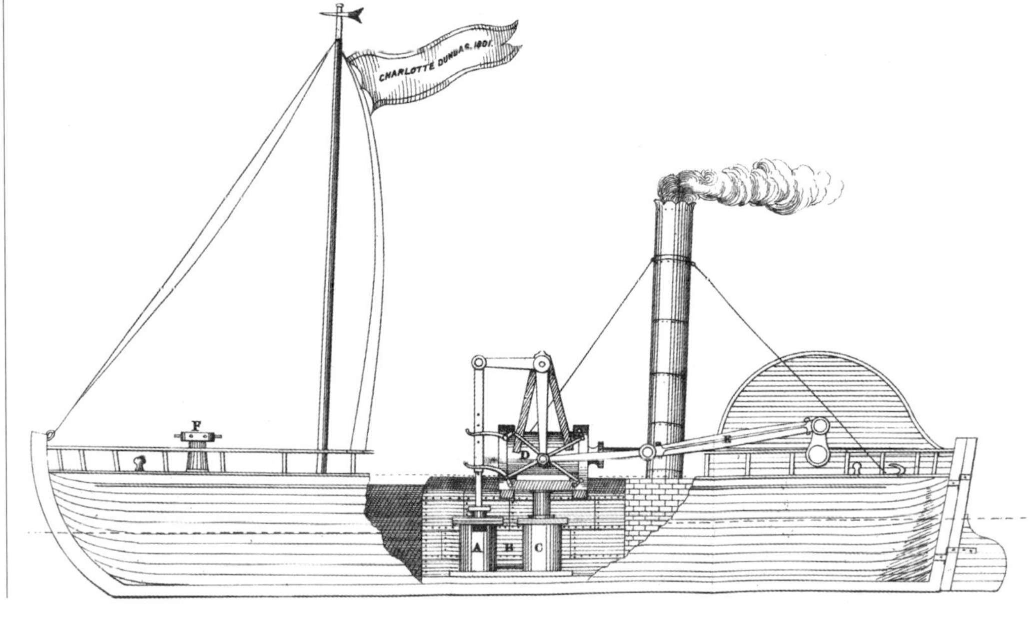

the onlookers were highly impressed – including the poet Robert Burns, who is said to have been on board to witness the feat.

The following year saw a more ambitious steamboat trial. Using Miller's influential position as a key shareholder and banker to the Carron Iron Company near Falkirk, Symington was able to construct a much larger vessel to be tried out on the Forth and Clyde Canal. The Carron Company spared no expense in building to Symington's design a twin-hulled paddle boat, sixty feet (eighteen metres) long, powered by a larger and more powerful engine. When the vessel was taken out for its first trial on 2 December, the engines proved too powerful for the paddle wheels, which began to disintegrate under the pressure. So Symington redesigned the wheels to make them stronger and, on Boxing Day 1789, the new ship surged along the canal at seven miles an hour.

All was not plain sailing after that, however. When Miller, a spendthrift, saw the bill for the new ship, he was appalled and abandoned the project. He was encouraged to give up on Symington by rival steam pioneer James Watt, who argued that Symington's engines were inefficient. Miller came to regard Symington as 'a vain, extravagant fool'.

Happily, however, Watt later made amends to Symington by employing him to write up his steam experiments.

In the meantime, Symington continued to refine his designs and, in 1801, was given a fresh commission by a new sponsor, the powerful Lord Dundas. The result was the famous paddle steamer *Charlotte Dundas*, which gave very impressive demonstrations, again on the Forth and Clyde Canal. The future of steam was assured. And as for paddle-wheeled vessels, Miller and Symington would no doubt have been pleased to see that the last remaining ocean-going paddle steamer in the world is also a Scot – the Glasgow-built *PS Waverley*.

THE PROPELLER SHIP
Robert Wilson, 1803–1882

No sooner had the paddle-steamer appeared on the world stage than another Scot was thinking up a way of bettering it. Such is the spirit of Scottish invention. His name was Robert Wilson, an engineer from a humble background in Dunbar, East Lothian, whose fisherman father was drowned when Wilson was just seven years old.

The sea that had claimed his father's life held a fascination for Wilson. He became an expert at sculling. This is the technique of propelling a boat using a long, flexible oar, mounted at the stern (rear) of the vessel. The oar is twisted from side to generate forward thrust. While working as a joiner's apprentice, Wilson began to sketch out how the twisting and turning action of the sculler's oar might be improved on. It is said that he was also inspired by watching windmills, which turned with a propeller-like action in the wind. He came up with the idea of a screw propeller, fixed to the stern of a boat, which was turned using a hand crank.

In 1827, Wilson showed a model of his propeller to the wealthy and well-connected Earl of Lauderdale. The Earl helped drum up interest in the propeller and, as a result, Wilson was awarded a medal by the Royal Scottish Society of Arts. His invention was presented to the Royal Navy, then known as the Admiralty, and it was given a trial. Although the trial was a success, it was considered too small-scale to be of use in the Navy. Wilson was keen to couple his propeller to a steam engine, but unlike many other inventors of the day, he was not from the kind of wealthy or well-connected background that would have made it possible to raise the necessary funds for such an expensive undertaking – not to mention find potential customers for the technology.

Others picked up on Wilson's invention, however. When the propeller was coupled to a steam engine, early trials showed that this system was faster, safer and more agile than a paddle wheel. In 1840, another inventor, Francis Petit Smith, demonstrated a two-hundred horsepower steam-driven propeller ship to the Admiralty. The authorities were impressed, and propeller ships went on to become the norm in the Navy, with merchant and pleasure craft following suit.

Although there was some debate as to exactly who invented the screw propeller, Wilson himself was unequivocal. In 1860 he published a pamphlet about his invention, entitled *The Screw Propeller: Who Invented It?* He took out more than twenty patents for propellers and other machinery to confirm the originality of his work, and to ensure he got his just reward. In 1880, he was paid £500 by the Admiralty to use his patented double-action screw propeller in a torpedo.

And Wilson was no one-trick pony. As manager of the Bridgewater Foundry near Manchester, owned by fellow inventor James Nasmyth, Wilson conceived many other innovations. These included an improved version of Nasmyth's steam hammer, a machine that bent iron pieces into useful components.

THE IRON-HULLED SHIP
William Fairbairn, 1789–1874

If you have been unlucky enough to be on a ferry or cruise liner during a severe storm at sea, you will know how unpleasant it can be. But at least you can be assured that the vessel is strong enough not to be broken up by the force of the ocean – no matter how much wave power is thrown at it.

This was not always the case. Although it might seem strange to think that water is strong enough to slice through metal, that's exactly what happened to the first iron-hulled vessels. Their poor design meant that many of them were easily split in two by the stress of sitting on the crest of a large wave while bearing a heavy cargo. So whole idea of a metal-hulled ship was almost sunk – until it was rescued by a structural engineer from Kelso by the name of William Fairbairn.

Fairbairn studied as a civil engineer and made his name by investigating how and why metal structures failed under pressure. With insights derived from his research, he designed bridges and other large structures for railways. His belief that iron bridge sections should be tubular to give them maximum strength led to a building

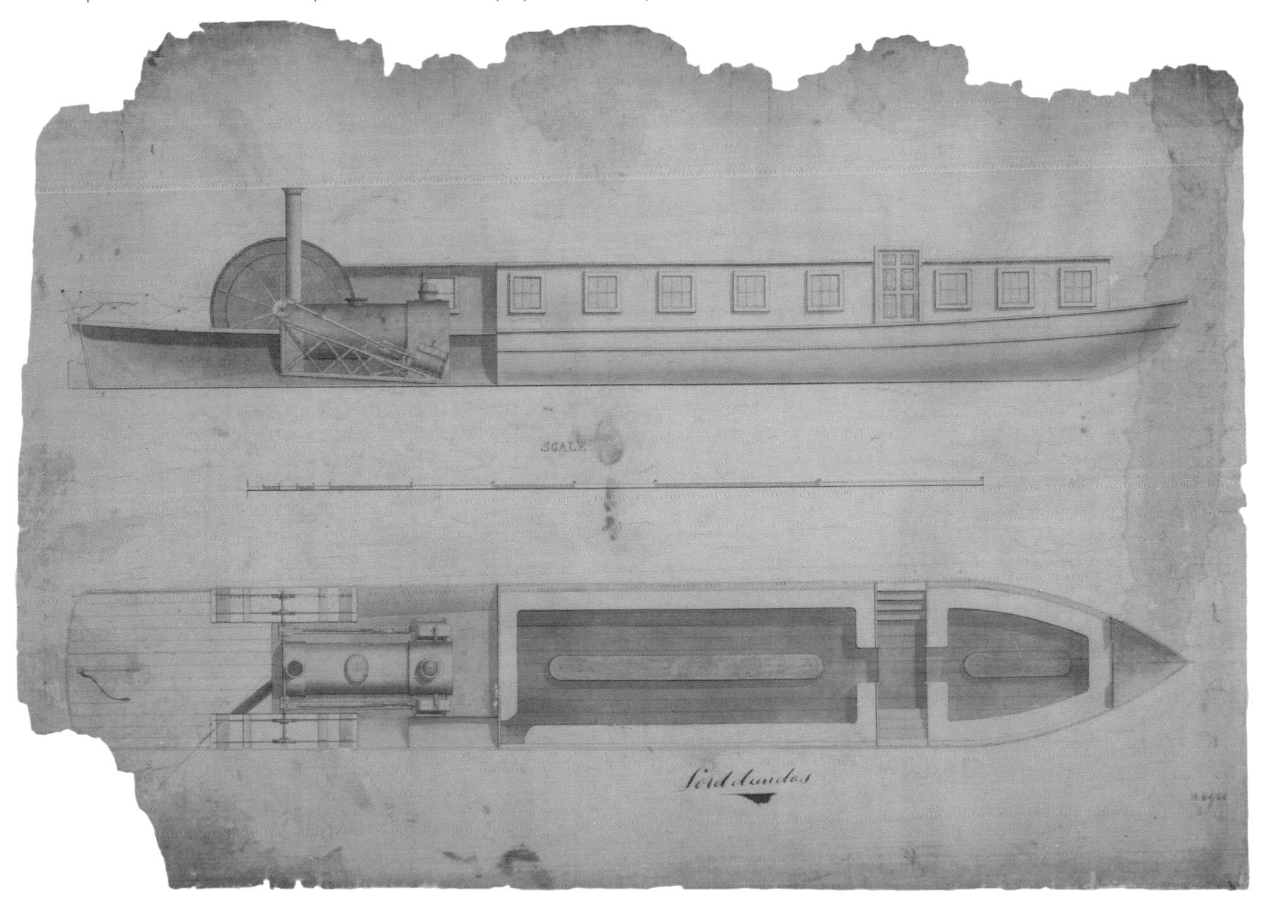

Lord Dundas, one of William Fairbairn's steamers.

revolution. His work helped prevent further disastrous and fatal bridge collapses of the type that had plagued early iron bridge designs.

In 1865 Fairbairn demonstrated how his ideas applied to the sea as much as to land by publishing his *Treatise on Iron Shipbuilding*. He argued that in order for iron ships to be tough enough to cope with the stresses and strains of life on the ocean waves, they had to be thought of as huge metal tubes. This meant the top of the ship had to be built as strong as the bottom of its hull, with extra strength built into the middle of the vessel to help prevent it buckling or snapping in two. Among the best known of Fairbairn's iron-hulled steamers were *Lord Dundas* and the *Megaera*.

To his list of innovations, Fairbairn added a new type of efficient steam boiler called the Lancashire boiler. However, it was his work on iron-hulled ships and civil engineering projects which had the most important legacy. In effect, Fairbairn partly invented the scientific discipline of structural engineering. It is from his investigations that later engineers acquired a fundamental understanding of metal fatigue and how it could lead to bridge and building collapses, boiler explosions, cracked ships' hulls and – many years later – aviation disasters such as those that befell the infamous Comet airliner.

THE LIFEJACKET
Captain Ward of the Royal National Lifeboat Institute, 1854

Should the worst happen at sea, there can be no better friend to hold on to than a lifejacket. The invention of the lifejacket is generally attributed to a Captain Ward, a Scotsman. Little is known about Ward other than that he was an inspector with the Royal National Lifeboat Institute (RNLI). In 1854 he invented a lifejacket for RNLI lifeboat crews, which made their work much safer since they often had to rescue other seamen in treacherous conditions and were in near constant danger of being drowned.

The design of Captain Ward's lifejacket was simple. Cork blocks were strapped together to form a vest, which the lifeboatman (there were no women in the service in those days) wore over his waterproof oilskins. Cork was used because it is a very buoyant material and Ward must have been aware that Norwegian fisherman had for years tied cork blocks to themselves to help prevent drowning should they be swept overboard into the ferocious North Sea. But Ward's was the first specially designed lifejacket, and its basic principles are still at work in modern inflatable lifejackets today.

THE STEAM ENGINE
James Watt, 1736–1819 and William Murdock, 1754–1839

If the steamship is an icon of travel, then another is the steam locomotive. Even today, in the age of diesel and electric trains, and even hover-trains, steam locomotives continue to ply some of the world's most scenic tourist routes. Apart from being the backbone of global travel in days gone by, what steam trains have always had in common with steam ships is the power plant at their heart. And, as every Scottish schoolchild should know, it was James Watt who invented the steam engine. Or did he?

The facts are that the steam engine developed over a very long time. It is known that a primitive steam-powered apparatus was in use in ancient Alexandria. Later, steam power was used in mediaeval Asia and Europe. These early dabblings in steam power were ingenious, however they could not really be described as engines. The same goes even for the steam-powered toy car design that was drawn in the 1670s by Ferdinand Verbiest, a Flemish missionary at the court of the Chinese Emperor. Verbiest's toy car was brilliant and its working

Left. An early life jacket.

Opposite. Diagram of Watt's steam engine.

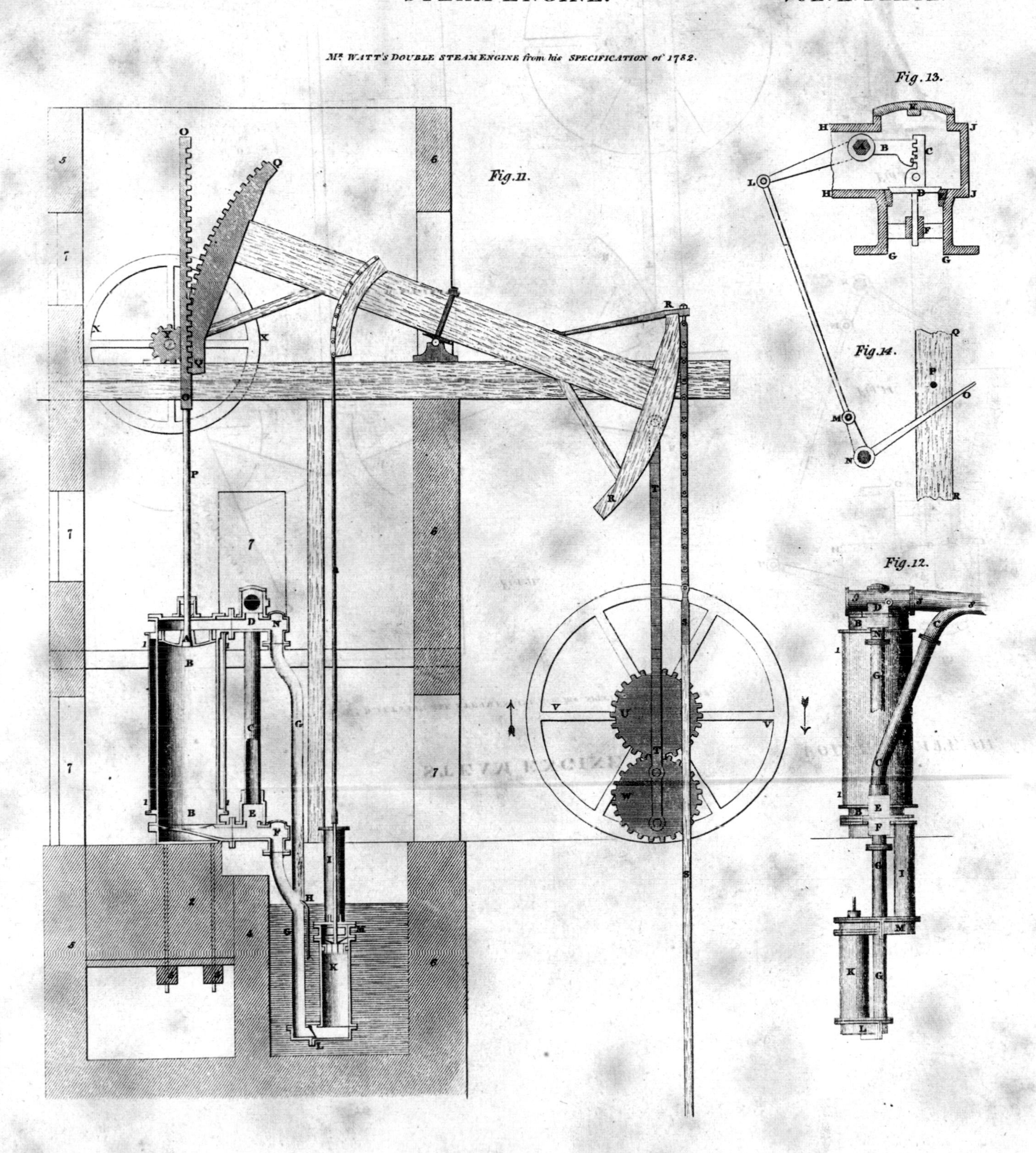

W. Creighton Del. Soho Staff.d

Engraved by W. & D. Lizars Edin.r

principles were sound. But there is no evidence it was ever actually built – and the technology would not have been powerful or safe enough to carry people, or power any serious machinery.

It was not until the early eighteenth century that an actual steam engine was invented, which could be used to power machinery efficiently and reliably. Devised by Englishman Thomas Newcomen in 1712, with input from some other inventors, the Newcomen engine was the world's first true steam engine.

So where does James Watt come in? Born in Greenock, near Glasgow, the young Watt got a job as an instrument repairer at Glasgow University in 1758. His enthusiasm and natural flair quickly brought him to the attention of one of the top university professors, the celebrated chemist Joseph Black. In his workshop Watt made repairs and conducted experiments with Black's encouragement. Watt was interested in steam power, and had been working on a model steam engine for some time when he got an opportunity to work on the small Newcomen engine that the University used to demonstrate steam power to students of natural philosophy. The engine was in need of repairs, so it was handed over to the eager Watt.

While examining the Newcomen engine, Watt identified some major flaws in the design that he reckoned he could put right – thereby making the engine more powerful, and much more efficient. The basic principle of steam power is this: water is heated in a chamber to produce steam, which creates outward pressure, because steam takes up more space than water. It is said that Watt first noticed the pressure of steam by watching it spouting furiously from a boiling kettle. In the steam engine, this steam pressure can be used to push up a piston connected to the chamber.

To pull the piston back down again, the pressure in the cylinder must be reversed by creating a vacuum. This is done by cooling the steam in the chamber to turn it back in to water, which takes up less space and so sucks the piston back in. The process is able to repeat itself automatically because the motion of the machine is harnessed to open and close valves which, for example, let cold water in to the chamber at the right moment to cool the hot steam. The result is a piston that goes up and down repeatedly, which can be used to pump water out of flooded mineshafts – as full-scale Newcomen engines generally did.

The trouble with the original Newcomen engine, Watt discovered, was that more than three-quarters of the engine's power was wasted inside the engine itself. This happened because the steam cylinder, which needed to be kept very hot in order to produce steam, kept losing its heat because cold water was sprayed directly into it in order to condense the steam back into water. Watt realised that what the engine needed was a separate condensing chamber, which could be kept cold, while the main steam chamber remained hot. In effect, Watt came up with an entirely new type of steam engine – faster, smoother and more efficient.

Watt built a fully developed version of his new machine in 1775. Its development and construction was paid for by Matthew Boulton, an enlightened and generous English entrepreneur. Boulton set Watt's engine to work in his new iron foundry in Birmingham. The machine was an instant success. Thanks to Watt's hard work and engineering genius, and Boulton's skill as a marketing man and business operator – he protected Watt's technology from piracy and struck sales and licensing deals that eventually made the two men very wealthy – the Boulton-Watt steam engine soon wiped the floor with the Newcomen engine.

Watt made continual improvements to his engine and, along the way, coined the terms 'Watt' and 'horsepower' as units to measure an engine's strength. The latter was a very useful advertising tool. 'How many horses,' a customer might ask, 'would it take to rival the power of Watt's engine?' 'A team of forty or more,' might come the reply. Followed by, 'Sign here, please!'

By the early 1800s Watt's engine had all but entirely replaced all other forms of large-scale power plants in mills, factories, foundries and mines across the UK and then the world. It also came to be used, in a modified mobile form, in steam ships and locomotives. But Watt did not achieve these further innovations alone. It was Boulton and Watt's young assistant, William Murdock, another Scot from Old Cumnock in Ayrshire, who turned Watt's stationary pumping engine into a mobile rotary engine for use in spinning machinery and vehicles.

Murdock, also known as Murdoch, realised that the up-down piston of a steam engine could be made to turn an axle – just as we can today observe how the up-down motion of a cyclist's leg pumps a pedal to drive the bike's rear wheel. The result was Murdock's 'sun and planet' drive system for rotary steam engines. This technology was invaluable for industry, but it had other applications, too.

In 1784 Murdock, who had been sent to Cornwall where he acted as an engineering supervisor for Boulton and Watt, built a working steam carriage. This was a robust, road-going steam locomotive, a prototype of the steam train. Unfortunately Watt became rather jealous of his young employee's brilliance, and Murdock's ambitions for the steam carriage were stifled. It was left to Murdock's friend and neighbour, English inventor Richard Trevithick, to develop the Scotsman's design further, successfully putting it on rails in 1804.

In the meantime, Murdock made further improvements to the steam engine that made it ideally suited for use in shipping. He lived long enough to witness his design power the *Great Western* steamship in the first steam-powered Atlantic crossing in 1838. As the stage was set for the golden age of steam, it became clear that the shared Scottish innovation of Watt and Murdock didn't just power the industrial revolution – it *was* the industrial revolution.

ROADS, BRIDGES, VIADUCTS AND CANALS

Telford, Rennie and Arrol, c.1800–c.1900

A tour of Scotland is a magical experience. Whether travelling by road, rail, sail or canal, sooner or later the visitor to Caledonia has to marvel at the many Georgian and Victorian-era roads, bridges, viaducts, harbours and canals that make their visit such a pleasure. The Scots who designed many of the earliest and most important of these types of structures, in Scotland and abroad, were also the men who really fathered the scientific discipline of modern construction, now known as civil engineering.

It was perhaps the mere fact of being born into such a beautiful yet rugged and unforgiving landscape that gave these Scots engineers their calling in life. Foremost among them was Thomas Telford (1757–1834). In his long career, Telford, who hailed from Langholm in Dumfries and Galloway, designed dozens of major bridges across the UK, most notably in the hilly terrains of Scotland and Wales. His Menai Suspension Bridge in Wales is one of the first of its type in the world, and pushed existing technology to the limit with its system of linked iron sections to form the suspension cables. He was also a brilliant road designer, improving on the designs of Macadam roads and earning himself the punning nickname 'Colossus of Roads', after the giant Colossus of Rhodes statue in ancient Greece.

What Telford did for bridges, John Rennie (1761–1821) did for harbours. Rennie, of East Linton near Edinburgh, designed some of the most famous docks and harbours in the UK, including Edinburgh's Leith Docks and the famous London Docks. In addition, he designed some huge projects for the Admiralty when Britain's military naval power was at its height, including the enormous breakwater at Plymouth Sound, used to protect Britain's Channel fleet from French attack. Yet Rennie, like Telford, was also a very accomplished bridge designer.

When it comes to bridges, however, there is one Scottish designer who arches over the rest. He was not as early or perhaps as pioneering as Telford or Rennie, but William Arrol (1838–1913) was a quite brilliant civil engineer. Born in

Telford's design for the Dean Bridge, Edinburgh.

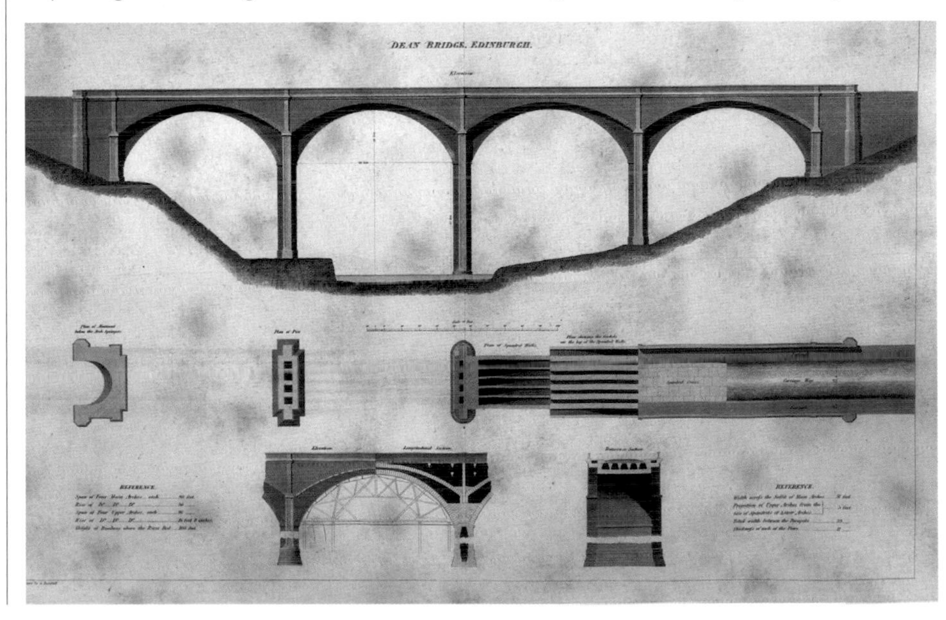

Thomas Telford.

Houston, Renfrewshire, into a very humble household, Arrol trained as a blacksmith as a boy while learning mechanics at night school. He went on to work at a Glasgow firm of bridge builders. But by the time he was in his early thirties he had set up his own company.

As much an astute and hard-working entrepreneur as a gifted engineer, Arrol used his biggest venture, Sir William Arrol & Co, to make himself a household name. It was through this company that Arrol designed the Tay Bridge (not the one that collapsed), Tower Bridge in London and the spectacular Forth Bridge – an awesome structure that took more than four and a half thousand men to build, almost a hundred of whom lost their lives in the process.

THE RAILPLANE MONORAIL
George Bennie, 1891–1957

How will we travel in the future? In 1930, one man thought he knew the answer. Years from now, reckoned Glasgow-born inventor George Bennie, people will go on holiday on the 'railplane'. Bennie came up with the idea for his railplane, a type of monorail, in the 1920s. In those days, futuristic modern inventions were all the rage. People believed that technology could make the impossible possible, and turn dreams into reality.

Luckily for Bennie, he came from a wealthy family, since his father ran a successful engineering firm. So, after serving in the First World War, Bennie didn't have to find paid employment. Supported by his family, he lived with his mother on the Isle of Bute and devoted his time to inventions.

Bennie didn't just sit around tinkering, however. He was a serious and hard-working inventor, and inherited a talent for engineering from his father. Bennie was golf-mad, and among his minor inventions was a new type of putter for use on the golf course. But it was his monorail that took up most of his time. The railplane monorail featured a carriage that was inspired by airship gondolas, which were hung beneath a vast air balloon and set in motion by propellers.

Unlike an airship gondola, the railplane carriage was slung from wheels that ran on a rail above. The rail was itself hung from a series of steel gantries, a bit like a long line of electricity pylons connected together. The carriage also had another set of wheels that ran along another stabilising rail below to prevent the carriage from swaying dangerously, since the electric-powered propellers at the end of each carriage were intended to thrust passengers forward at speeds of two hundred miles per hour. According to Bennie, his railplane was as close as you could get to flying without actually taking off.

In 1927 Bennie set up a company to build and market his invention. A prototype test track was built at Milngavie train station just outside Glasgow. Many of the great and good, politicians and businessmen, came to try out the railplane for themselves. It was impressive, but unfortunately for Bennie, it was also deemed too expensive. It was thought that conventional trains were fast enough

George Bennie's Railplane.

and more cost effective, especially in light of the Great Depression that followed the 1929 Wall Street stock-market crash.

Bennie, who remained a bachelor all his days, continued to pour money into his project, trying to drum up interest. But it was to no avail. He was declared bankrupt in 1937 and lived just long enough to see his Milngavie prototype dismantled and sold for scrap in 1957. Although Bennie's propeller-powered railplane did not catch on, other monorail systems did and are in use today in cities across the world – with some of them reaching speeds of more than three hundred miles per hour.

PICTURE POSTCARDS
George Stewart & Co, 1894

Wish you were here: the classic postcard greeting from a friend on holiday at the seaside still radiates warmth, more than a century after the postcard was invented. But often it is the picture, especially when the card has been thoughtfully chosen, that really raises a smile – or even a giggle. And once again, the colourful picture postcard is another classic Caledonian invention. Who said the Scots were dour?

When postcards first appeared, they did not actually have a picture on them. These cards were introduced by the postal service in Germany in 1865 as a cheaper alternative to letters or telegrams. Five years later the UK Post Office followed suit. On one side was a pre-printed halfpenny stamp to pay for postage, and a space for you to write the recipient's address. The other side was for your message.

The early Post Office postcard was basic, but very popular. In those days, of course, there was no email, no mobile-phone texts nor even traditional telephones. So unless you had something terribly private to say, the postcard was the ideal way to dash off a message to your loved ones that you were having a whale of a time in Wales – no doubt having made use of various Scottish innovations such as steam power and railway bridges in order to get there.

To begin with, the Post Office didn't allow anyone else to make postcards in the UK, but, in 1894, it was persuaded to change its mind. Seizing on the opportunity, the Edinburgh firm of George Stewart & Co is believed to have been the first private stationery firm to print illustrated postcards. A halfpenny adhesive stamp (another Scottish invention) was stuck to the front alongside the address, with the picture and the message occupying the other side of the card.

In 1902 the Post Office permitted the stamp, address and message all to be put on one side of the card – freeing up the whole of the other side for pictures. These ranged from Stewart's local landmarks and scenic views of the 1890s to the famously saucy seaside picture postcards of the 1930s.

PARAFFIN
James Young, 1811–1883

What better way to enjoy the great outdoors than on a camping holiday? But when the rain starts pouring down, as it so frequently does in Bonnie Scotland, you'll be glad you packed your stove for cooking the dinner on. Nowadays, portable camping stoves often use gas as a fuel, but traditionally paraffin – a type of oil – was often used. And commercially available paraffin, along with many other types of cooking, heating and fuel oils, as well as gas, owes its origins to a undertaker's son from Glasgow named James Young.

In effect, Young invented the oil industry. He started out by attending evening classes in chemistry at the Andersonian Institute, now Strathclyde University. He was a very bright student and, after graduating, progressed to a career in the chemicals industry. A friend from university days told him about a coal mine in Derbyshire which contained an oil well. When Young experimented with the oil, he found he could refine it to create a number of usable oils such as paraffin, lubricating oil and naphtha, an industrial solvent.

At first, Young thought the oil was coming from the coal in the Derbyshire mine. So he asked friends and acquaintances from all over the UK to send him samples of coal found in their local area in the hope that he would find another similar oil well somewhere else, and start a business using it. He was particularly intrigued by one parcel of coal from Bathgate, East Lothian, in his native Scotland. The substance in the parcel was described as 'cannel coal' because, when lit, it burned as brightly as a candle. In fact, the substances in Derbyshire and East Lothian were not coal at all, but oil shale – an oily sedimentary rock which, when heated under certain conditions, produces usable gas and oil.

In 1850, Young patented a process for refining oil and gas. He extended his patent to North America, so that the first American

oil tycoons had to pay him a royalty every time they wanted to use his technique. Back home, he set up the world's first oil refinery on the outskirts of Bathgate, where rich underground shale seams were located. After Young's Paraffin Light and Mineral Oil Company was set up, an oil craze swept the UK. By 1870 almost a hundred oil companies were operating in West Lothian alone, employing thousands of workers and making such products as paraffin and Scotch Petrol.

Young had established a shale oil industry that lasted until the early 1960s, by which time the cheaper costs and easier production of mineral crude oil had won over producers. The Scottish oil industry was soon to be revived in a new guise thanks to the discovery of mineral oil and gas deposits under the North Sea. Young – who had also come up with a proposal for a Channel tunnel years ahead of its time, as well as financing the African expeditions of his friend Sir David Livingstone – continues to have a lasting and very visible legacy.

Apart from laying the foundations of the vast and often avaricious global petroleum and petrochemicals industry – without which none of us would get very far on holiday, whether by boat, plane, train or automobile – plus various generous donations from his vast personal fortune, Young also left a gentler monument to his work. That monument is the shale bings that can be seen from both the M8 Edinburgh-Glasgow motorway and the Edinburgh-Falkirk-Glasgow rail line. These bings are man-made hills, created from by-products of the vast shale extraction works of Young and his followers in the nineteenth and early twentieth centuries. And, as proof that oil exploitation is not without any environmental merit, these organically rich bings now provide a nurturing habitat for a range of wild plants and animals.

THE THERMOS FLASK
James Dewar, 1847–1932

The Thermos flask is surely an indispensable part of any weekend camping holiday or country hike. In the cooler months of the year it keeps hot drinks hot, and in the sweltering summer keeps cold drinks cold. Such an elegantly simple premise, and again the work of a Scottish inventor. Although the household name of Thermos belongs to the German company that marketed the first commercial flask in 1904, it was designed by the physicist and chemist James Dewar in the early 1890s.

Dewar, who was born in the small Fifeshire town of Kincardine-on-Forth, came up with the idea for the flask while experimenting with ways to store liquid gases. Such liquid gases as nitrogen have to be stored at an extremely low temperature – 197 degrees Celsius below freezing – to prevent them boiling and hence evaporating. So Dewar came up with the idea of a vacuum jacket for the vessel storing the liquid gas.

The air between the inner vessel and the jacket was extracted, creating a vacuum, so that heat could not be transferred via conduction or convection between the inner and outer skins of the flask. In addition, the walls of the flask were coated in silver to further reduce any heat loss by radiation. The Dewar Flask worked so effectively in the long-term storage of liquefied gases that it wasn't long before others recognised its potential for storing drinks – and the Thermos was born.

NATIONAL PARKS
John Muir, 1838–1914

In 1875, keen mountaineer John Muir was trapped for four days in a blizzard at the top of a mountain in California's Sierra Nevada. The freezing storm that made it impossible for Muir to descend also gave him frostbite, permanently damaging his feet. Eventually he made it down, starving

John Muir.

and exhausted, and vowed he would do it all again in a heartbeat. Nothing could dampen the enthusiasm of the man who loved the great outdoors so much he invented environmentalism, the animal-rights movement and the National Park.

While most of us would prefer not to go to such extreme lengths as Muir to find excitement and adventure while on holiday, we are the beneficiaries of his lifelong love affair with nature. Before his family emigrated from Scotland to America, the young Muir loved to roam the countryside of his native East Lothian. When he was eleven, he crossed the Atlantic and helped his father clear a patch of forest to create a family smallholding in Wisconsin.

From his stern and deeply religious father Muir inherited a Biblical view of the world. In 1867, he was blinded by an accident while working as a mechanic. His sight gradually recovered, and as it did so he decided to go an a pilgrimage across America to discover its natural treasures. Along the way Muir discovered the Yosemite region and the Sierra Nevada mountains, which left a deep and lasting impression on him.

Muir settled in San Francisco, California, within relatively easy reach of the Sierra Nevada. He grew concerned at the destruction of the wilderness by industry and commercial farming. So he set about lobbying for Yosemite to be preserved for future generations as a National Park. He did so with the help of his friend Robert Underwood Johnson, editor of the magazine *Century*, who commissioned Muir to write a number of articles about nature and its conservation.

Eventually, in 1890, Muir and Johnson's efforts paid off, and Yosemite was declared a National Park, covering an area of more than a thousand square miles. Two years later Muir set up the first environmentalist group, the Sierra Club, to encourage people to visit and enjoy America's wild places – but also to try to protect those places, and the animals and plants within, from the forces of commercialisation. He made numerous studies of the ecology of such places as Yosemite and wrote them up as articles in various publications.

In 1903 Muir took US President Theodore Roosevelt on a camping trip deep into the Yosemite Park, and along the way convinced Roosevelt to give the park federal government protection. Years later, the elderly Muir was horrified when Roosevelt's successor, Woodrow Wilson, approved the flooding of the stunningly beautiful Hetch Hetchy Valley to create a water reservoir for San Francisco.

However, Muir's opposition to the Hetch Hetchy scheme did at least help to make the cause of environmentalism front-page news. After his death, Muir became an American national hero, especially in California, where his birthday of 21 April is celebrated as John Muir Day. The National Parks movement that he began spread across the world. Today there are hundreds of such parks in every corner of the globe. In Scotland, Muir's work is continued by the John Muir Trust. He even has an asteroid named after him.

HOLLOW-PIPE DRAINAGE
Hugh Dalrymple, 1700–1753

It doesn't sound very exciting, but good drainage is one of the foundations of modern civilisation. Without it, your toilet would get blocked, your roof would leak and the road to your house would be impassable – certainly in a country with as much rain as Scotland.

Drainage first became a racy subject in the eighteenth century. A lot of wealthy and well-educated farmers began looking at the bits of their estates that weren't useful for very much, and asking questions like: why isn't it useful? And what can I do to improve it? For example, how can I turn that waterlogged scruff of land near the river into a field fit for cultivation?

Before Hugh Dalrymple came along the answer would have been: 'You can't'. But Dalrymple, who was a member of the powerful Dalrymple of Drummore dynasty in East Lothian, came up with a new bit of kit to change all that. He kept his invention simple but highly effective. It was a hollow pipe, laid in a trench in the ground and then covered over. Through it, moisture was drained out of the soil and into a ditch or reservoir.

Result: a nice, dried-out field for growing crops in. Dalrymple's invention was a classic example of a period in history known as the Scottish Enlightenment – when a dizzying array of bright ideas were first put into practice to improve our way of life, including many of those revealed in these pages. Like the Scotch plough, for example . . .

SCOTCH PLOUGH
James Anderson, 1739–1808

You're fifteen years old. Your parents have just died. There's a large family farm relying on you to keep it going. What do you do? If you're James Anderson of Hermiston, near Edinburgh, you swallow your grief and then roll your sleeves up and get on with it.

Anderson was a very bright and hard-working young man. So

EXCAVATORS AND BULLDOZERS

James Porteous, 1848–1922

Here we have an invention that made the earth move. James Porteous – another enquiring mind from East Lothian, and another millwright – came up with the machine from which all modern excavators (or earthmovers) and bulldozers are at least in part descended. He did so after emigrating from Scotland to the United States, settling eventually on the west coast – in Fresno, in California's San Joaquin Valley. There, Porteous used the engineering and farming skills he had learned from his father to set up the Fresno Agricultural Works.

Porteous's machine, unveiled in 1883, was called the Fresno Scraper. It came about because Porteous realised that agriculture in the valley could be greatly improved if there were many more ditches and canals to irrigate its dry, sandy soil. The traditional way of making such improvements was for teams of men to shovel earth into horse-drawn carts. But Porteous wanted to devise a means of making earth-moving much more rapid. Porteous was aware of some existing designs for horse-drawn earth scrapers. So he bought the patents for these and then added key innovations of his own to produce an entirely new and truly 'groundbreaking' machine.

Instead of simply pushing earth along, as earlier machines did, the Fresno Scraper had a blade attached to the bottom of a C-shaped bucket, which scooped the earth up. But the really clever bit was the fact that the bucket could be tipped, so that soil scooped up from one place could be moved, and then tipped out somewhere else. It proved to be at least four times faster than existing ways of working.

At first, the Fresno Scraper was mounted on runners and dragged by horses. Over time the runners gave way to wheels and the horses gave way to diesel engines. Eventually, by the time Porteous died in 1922, modern bulldozers and excavators were evolving from both his design and the steam shovel of William Otis. But not before the ground had been cleared by thousands of Fresno Scrapers – including their use in the construction of the famous Panama Canal, which joined the world's biggest oceans, and in the trenches of the First World War.

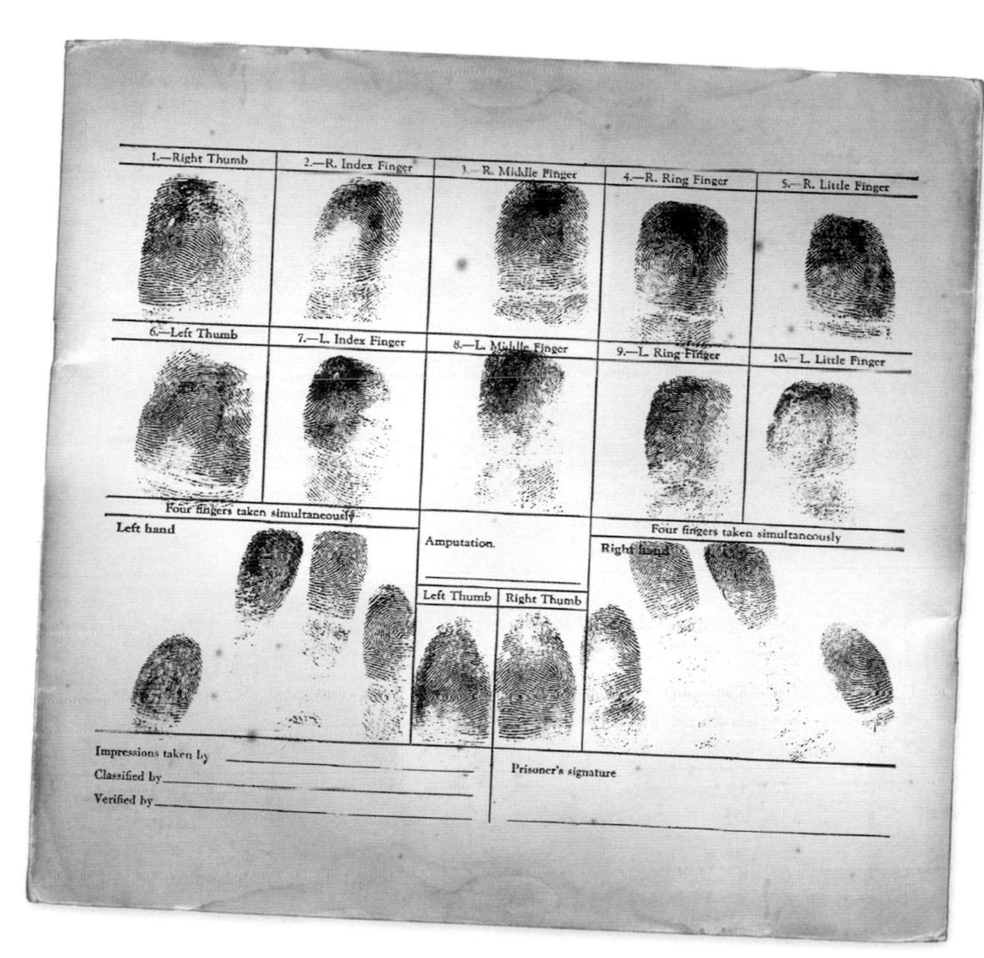

1.—Right Thumb	2.—R. Index Finger	3.—R. Middle Finger	4.—R. Ring Finger	5.—R. Little Finger
6.—Left Thumb	7.—L. Index Finger	8.—L. Middle Finger	9.—L. Ring Finger	10.—L. Little Finger

Four fingers taken simultaneously
Left hand

Amputation

Left Thumb | Right Thumb

Four fingers taken simultaneously
Right hand

Impressions taken by
Classified by
Verified by

Prisoner's signature

CRIMINAL FINGERPRINTING
Henry Faulds, 1843–1930

In 1892, two young boys were brutally killed in a village in Argentina. At first, the police suspected the wrong person. But the case was eventually solved thanks to the identification of a fingerprint left by the murderer – the boys' mother – at the crime scene. It was the first time fingerprints were successfully used in a murder investigation. A few years later, fingerprints made their first appearance as a clue in a Sherlock Holmes detective story – 'The Adventure of the Norwood Builder'. Since then, fingerprinting has become legendary as a method of identifying criminals.

The way fingerprinting works is ingenious. Prints found at a crime scene – on a body, a knife, a wine glass or a door handle for example – are photographed. Sometimes these fingerprints are not patent (visible, as in bloody or dirty) but latent (invisible) so a special dust or other compound must be used to make them visible before they can be photographed.

Investigators then compare these prints with fingerprints they have taken from suspects at the police station. Suspects' prints are obtained by pressing their fingers on an ink-pad and then pressing them against a card. The resulting mark is photographed and laid next to the photograph of the prints found at the crime scene. If the two match exactly, then the culprit has been identified – because no two human fingerprints are alike.

So who came up with this amazing invention? Fingerprints were used as signatures in medieval societies, but people did not really understand that every fingerprint was unique – far less did it occur to anyone to devise a system of criminal fingerprinting. That honour goes to a Scottish doctor called Henry Faulds.

Faulds was born in 1843 in the village of Beith in North Ayrshire. After studying medicine he became a medical missionary, his job to help improve the health of people across the world. In 1873 he was sent to Japan, where he became a fluent Japanese speaker and founded a hospital in Tokyo.

It was while Faulds was in Japan that he first became interested in fingerprints. In his spare time he attended archaeological digs, where ancient Japanese pottery and other treasures were excavated and examined. He noticed that on some of the pots, the fingerprints of the people who made them remained visible. They had left an impression while the clay was still wet, which remained preserved forever after the clay dried and hardened.

It occurred to Faulds that, in everyday life, grimy fingertips often left marks on the surfaces they touched. He also realised that, when examined closely, the fingerprints left by one person were different from those of another. This meant you could identify an individual by the unique 'signature' left by the whirls and ridges on their fingertips.

In 1880, Faulds published an article about fingerprints in *Nature* magazine. In his article, Faulds suggested how fingerprints could be used to catch criminals. He then spent six years devising a system of forensic criminal fingerprinting that could be used by police investigators. But when he returned to Britain and presented his system to Scotland Yard in London, the police declined to take it up.

It was not until 1892, when the police inspector investigating the murder of the Argentinian children decided to test for fingerprints, that the system was widely adopted. Faulds meanwhile became embroiled in a dispute with other scientists over who actually came up with the forensic fingerprinting system first. Although it was eventually conceded that Faulds was the original inventor, until the day he died he remained bitter about the lack of recognition he received.

THE PRIVATE DETECTIVE
Allan Pinkerton, 1819–1884

Ever get the feeling you're being watched? Well, if you worked on America's railroads in the 1800s, and were regarded as a troublemaker by your bosses, then the chances are you probably were. And the folk doing the watching were likely to have been agents from the world's first private detective agency – Pinkerton's. The agency was set up in 1855 by Allan Pinkerton, a Scotsman from Glasgow.

Pinkerton is probably the most famous real-life spy in history. He created an insignia for his company that consisted of an image of a large, watchful human eye above the slogan 'We never sleep'. It's perhaps no surprise, then, that Pinkerton also came up with term 'Private Eye' to

describe his work. He also published the first detective stories, based on the work of his agency. These best-selling books were to be a big influence on Arthur Conan Doyle and his Sherlock Holmes stories, as well as later detective-story writers such as Raymond Chandler.

The difference is that Pinkerton's adventures were real. He was brought up in poverty, because his policeman father was unable to work due to injuries sustained on the job. Pinkerton must have felt that this was a very unfair situation, and it seems to have had a big influence on his life. He became an apprentice cooper – someone who makes wooden barrels – and joined the Chartists. The Chartists campaigned for a fairer society, with better working conditions for the poor and the right to vote for all men. In those days, only a very small number of wealthy people had the right to vote. Unfortunately for Pinkerton, Chartism was outlawed and he soon got into hot water with the authorities.

Pinkerton decided to emigrate for a new life in Canada, and took his newly wedded wife with him. The couple were shipwrecked off Nova Scotia, but survived and made their way across the border to America in search of work. They arrived at the immigrant town of Dundee north of

Chicago, where Pinkerton set up his own barrel-making business. He never forgot his radical roots, and in his spare time he helped American slaves escape across the border to Canada.

The idea of becoming a detective came to Pinkerton after a brush with some criminals out in the woods. While he was looking for logs to make barrels with, Pinkerton stumbled upon the secret hideout of a gang making counterfeit money. He tipped off the local police, who arrested the crooks. The authorities were impressed. It quickly became clear that Pinkerton had a talent for detective work and, after he helped the local sheriff catch even more criminals, he was given the job of deputy sheriff.

In 1849 Pinkerton was Chicago's first full-time police detective. Being as ambitious and cunning as he was tough and fearless, however, Pinkerton soon realised that he could make more money if he set himself up as a private detective who contracted his services to the authorities. He was right. The Pinkerton Detective Agency went on to solve a series of train robberies, which made his agency famous. Along the way, Pinkerton invented or perfected a number of classic techniques of the private dick. These included shadowing, or tailing, suspects, staking-out premises and going undercover to infiltrate criminal gangs.

Pinkerton's growing reputation impressed the controversial President-elect of the United States, Abraham Lincoln. In 1861, Pinkerton was hired by Lincoln to act as his personal bodyguard. It was just as well Lincoln did so, since Pinkerton foiled a plot to assassinate the president en route to his inauguration ceremony. Pinkerton was rewarded by being made head of the president's secret service. When the American Civil War broke out between the Union states on one side and the Confederates on the other, Pinkerton helped the Union side by going undercover as a Confederate officer, using a fake name, to gather

military intelligence.

After the war, Pinkerton returned to running his private detective agency. It became famous for its professionalism, painstaking research and high success rate – but also notorious for its often brutal and callous methods. Just as Pinkerton popularised the use of photographs or 'mug shots' of suspects to help solve crimes, his agency also came to rely on the use of intimidation, bribery, violence and even murder.

The Pinkerton Agency's underhand tactics were at their worst after Pinkerton handed control over to his two sons. Dirty tricks were often used to destroy the trade unionists who were fighting for fairer wages for workers and thereby threatening the profits of some of Pinkerton's wealthy clients – clients such as the Scots-American business tycoon Andrew Carnegie. These black marks against the agency – whose members came to be known pejoratively as 'Pinkos' – are especially saddening when we remember that Pinkerton had started out in life as a champion of better living and working conditions for poorer people and slaves.

In later years, Pinkerton turned his attention to writing books. In 1884, he suffered a freak accident, slipping on a Chicago pavement and biting his tongue. When the wound became infected, Pinkerton refused to have it treated, and it led to his death from blood poisoning. It was a bizarre and unexpected end for a man who had confronted death and danger so many times in his career. The Scot's name remains well known, thanks to his induction into the Military Intelligence Hall of Fame in the United States, where his Civil War agency work is also acknowledged as the forerunner of the modern secret services – namely the FBI, CIA and Department of Homeland Security.

THE MACINTOSH RAINCOAT
Charles Macintosh, 1766–1843

Thanks in part to an ex-employee of Pinkerton's Detective Agency by the name of Dashiel Hammett, this next Scottish invention has become world famous. We are talking about the Macintosh (or Mackintosh) raincoat. In the 1941 film version of Hammett's celebrated thriller story *The Maltese Falcon*, hard-boiled detective Sam Spade wears an overcoat that carries all the hallmarks of a classic Macintosh. Who can ever forget seeing the iconic image of actor Humphrey Bogart as Spade – with his coat collar turned up and his fedora hat pulled down over his brow?

The Macintosh no doubt became popular with real-life detectives because they spent a lot of time hanging around on street corners. They needed a coat that would keep them warm, dry – and yet coolly stylish – on a cold, rainy night. They needed a 'Mack'. The trademark of the classic Mack is a layer of waterproof rubberised lining devised by its inventor, Charles Macintosh, around two centuries ago.

Macintosh was born in Glasgow long before Spade or Raymond Chandler's Philip Marlowe were conceived. He was the son of a chemical manufacturer. Following in his father's footsteps, Macintosh made chemicals used to dye fabrics. Some of his chemicals, such as potassium and ammonia, he extracted from seaweed and human urine – the latter collected from barrels left on street corners. How's that for Scottish thrift and ingenuity?

In the early 1820s, Macintosh's experiments with chemicals led to the discovery that he could bond rubber with fabric using a chemical solvent called naphtha. He used his discovery to invent a waterproof fabric, which contained a layer of rubber sandwiched between layers of cotton or wool. He patented his invention in 1823. The following year Macintosh was commissioned to use his new invention to provide waterproof coats, lifejackets and airbeds for an Arctic expedition. This high-profile publicity led to a take-off in business.

In the years that followed, Macintosh at first supplied fabric for specialist tailors to make into garments. But he later moved into manufacturing his own ready-to-wear Macintosh coats after teaming up with Manchester clothes maker Thomas Hancock. As Hancock observed, early Macks were far from perfect. They could feel stiff and unpleasantly rubbery, with a tendency to smell and even melt in hot weather. Hancock solved many

of these problems by patenting a process called Vulcanisation, which improved the rubber element in the coats.

The Mackintosh brand went from strength to strength (picking up the letter 'k' along the way). In 1925, the firm was bought by Dunlop – another great Scottish company specialising in rubber – and today the genuine Mack is a sought-after garment, manufactured in Cumbernauld and sold to order. And should Sam Spade care to investigate, he will find that his trusty coat is still made, using a refined version of Macintosh's original process.

SPECIAL FORCES
David Stirling 1915–1990

Who Dares Wins. This is the regimental motto of the Special Air Service, or SAS, a British Army corps formed in the Middle East during the Second World War. It has since become the model for elite Special Forces throughout the world.

The SAS today consists of three highly specialised infantry regiments. Originally, however, it was devised as a single small-scale desert raiding party. Its inventor was an inexperienced but highly persuasive young subaltern officer called David Stirling.

Stirling, who was born in Keir, Stirlingshire, was a son of the Scottish gentry. His father was a brigadier-general in the Scots Guards and his mother was a daughter of Simon Fraser, Fifteenth Lord Lovat. The young Stirling was a happy and very privileged child. He demonstrated his yearning for adventure and danger when he abandoned his degree at Cambridge in order to go into training to climb Mount Everest. When war broke out in 1939, Stirling cut short his mountaineering practice in the Canadian Rockies and returned home to join the Scots Guards.

In 1941, Stirling volunteered for a new force called No.8 Guards Commando, which he rightly anticipated would provide him with greater excitement than soldiering in an ordinary regiment. The commandos were sent to North Africa, where Stirling formed a clear-sighted overview of how the war was going in the desert. The British and their allies were locked in a long, hot and seemingly interminable slogging match with the German Axis powers, with little sign of victory for either side. This gave the twenty-five-year-old Stirling an idea.

An audacious kind of guerrilla warfare was what Stirling had in mind, and was at the very least subconsciously inspired by his Scottish military heritage. A few decades earlier, Stirling's cousin, Simon Fraser, Fourteenth Lord Lovat, had formed a unit of brave and elusive Scottish snipers called the Lovat Scouts, and in some of the ways they operated they foreshadowed Stirling's plans.

Stirling had also imbibed a long Scottish martial tradition of raiding, guerrilla fighting and generally moving covertly about in small numbers to catch a larger and numerically superior enemy unawares. It was a tradition carried on down the centuries by the likes of King Robert the Bruce, who used guerrilla tactics against the English army during their occupation of Scotland in the 1300s, and by cattle-rustler Rob Roy Macgregor.

Stirling proposed to lead a well-equipped raiding party through the Sahara Desert, which, according to received wisdom, was an impassable sea of sand. Once on the other side, he and his raiders would be well behind enemy lines. This put them in a perfect position to strike at targets both valuable and vulnerable.

In characteristic style, Stirling took his idea straight to the top. He sneaked into the office of the Commander-in-Chief of the allied Middle East forces and persuaded him to grant authority to carry out the plan. Six officers and sixty other men were duly recruited to Stirling's force, which was given the deliberately misleading name of L-detachment Special Air Service (SAS) brigade.

Stirling's first operation, carried out in November 1941 by parachute, was a disaster, and the force suffered heavy casualties. But Stirling was undeterred. His next raids were by land, and these, with the invaluable help of another force called the Long-range Desert Group, were spectacularly successful.

Stirling and his men used simple but effective tactics. Use darkness as a cover; bluff your way past guards using your language skills; know how to find food and shelter in hostile territory; get in and out as quickly and quietly as possible, and so on.

In this way, hundreds of enemy aircraft were mined and destroyed on the ground. Fuel dumps were set alight and bridges and roads blown apart. Perhaps most importantly, enemy morale was sapped and their war strategy was undermined. Stirling was rapidly promoted up through the ranks and was a lieutenant-colonel by

the time he was eventually captured by the Germans in Tunisia in January 1943. He escaped four times but was recaptured and finally held in Colditz Castle until the fall of Hitler's Germany.

The SAS was disbanded at the end of the War. But it was soon realised that such a clever military invention would be of ongoing benefit to Britain's national security. So the SAS was reformed again in 1947 as one regular and two territorial regiments, with Stirling again at its heart. In the years that followed, SAS operations in such theatres of war as Malaya, Northern Ireland and the Falklands became well known and sometimes controversial.

Among the most legendary of the force's exploits occurred during six days in the late spring of 1980. Operation Nimrod, as it was known, vividly demonstrated the anatomy of a swift, audacious, high-tech and meticulously planned SAS strike. Using the latest electronic surveillance technology and weaponry, SAS troops stormed the Iranian embassy in London to free more than two dozen hostages who were being held by an Arab paramilitary group.

The operation, which was widely regarded as a spectacular success for the SAS, was staged in front of cameras from the world's media. It was a great publicity coup for the new Conservative government of Margaret Thatcher and turned the fiction of high-stakes, James Bond-style counter-terrorism into an apparent reality. In the decades since the sensational Operation Nimrod, a public hungry for heroes has devoured numerous fiction and non-fiction books, films and computer games inspired by SAS activities.

As the years passed, Stirling himself moved into other related fields, including international mercenary work and domestic political espionage in Britain. Yet he remained in constant contact with the SAS until his death in 1990. The elite fighting force he invented has continued to operate in 'deep space' combat in the hostile territories of Iraq and Afghanistan, as well as on the streets of Europe and elsewhere.

Over the years, the SAS has grown in fame. But it has also made costly errors and been engaged in some questionable operations. However, supporters of the SAS argue that in times of crisis the only alternative to taking action is to do nothing. This means never running the risk of making a mistake, but it also means leaving the United Kingdom and its people open to far worse consequences.

CAMOUFLAGE
The Lovat Scouts' ghillie suit, c.1900

A band of hardy and stealthy sharp-shooting Scotsmen, the Lovat Scouts were in many ways a precursor to the SAS. Formed during the Boer War of 1899–1902, the Lovat Scouts relied extensively on their secret weapon. And it was the ultimate 'secret' weapon – camouflage. The word comes from the French term *camoufler*, meaning 'to disguise'.

Camouflage of a sort was used by armies before the Lovat Scouts appeared, but these early examples used only basic methods such as painting uniforms in dull greys or browns to try to blend in with the environment. But with real camouflage, of the sort introduced by the Lovat Scouts, soldiers do not merely blend in with the environment. They become completely invisible.

Like insects or fish whose

The ghillie suit receives a royal reception.

bodies fool predators and prey by looking exactly like twigs or stones, the Lovat Scouts wore an outfit that looked exactly like the world they were fighting in. That outfit was the ghillie suit. Ghillie is the Scots Gaelic word for a 'servant' or 'guide' or 'boy', which in turn comes from the Old Norse word 'gille'. In the medieval Scottish Highlands, ghillies were expert hunters who assisted their lords and masters on expeditions and, more often than not, did most of the real hunting on their master's behalf.

As gamekeepers, ghillies also guarded the laird's deer and game from poachers. To perform this function, in particular, the ghillie needed to operate covertly and be invisible to the human eye. This is where the ghillie suit came in. Its precise origins are unclear, but the ghillie suit may have originally been an adaptation of the Highland big kilt, or *feilidh mhor*, which was made from a huge, five-metre piece of folded and pleated woollen cloth, often earthy in colour and designed to cloak the wearer while prowling on the moors.

Certainly, the ghillie suit was well developed by the nineteenth century. By then it might typically have consisted of a webbed garment made from twine into which were woven loose strips of cloth, leaves, twigs, heather and even clumps of turf. When made and worn by experts, the suit completely disrupted the appearance of the human form and made the ghillie an indistinguishable part of the Highland hillside. It was heavy, but then if the ghillie was well positioned, he didn't need to move much.

The potential of the ghillie suit and the men who wore them was recognised by Simon Fraser, Fourteenth Lord Lovat, an army officer and major landowner from Inverness-shire. After the outbreak of the Boer War in South Africa, Lovat approached the War Office in late 1899 with plans to form two companies of Lovat Scouts.

The Lovat Scouts were to specialise in stalking, spotting, sniping and riding, with many recruited from the traditional ghillie heartlands of the Highlands. The goal was to use the Scouts to weaken Britain's enemies in the conflict, the Boers, through superior tactics and with the aid of a certain invisible garment. The goal was achieved.

The ghillie suit proved its worth again in the World Wars. During the latter stages of the Second World War the Lovat Scouts were deployed in the mountains of Austria, where their hillcraft, mountain-warfare skills and unique camouflage enabled them to hunt down fleeing Nazi Party members. After the war, the Lovat Scouts were stood down. Many former members were then eagerly recruited by such elite fighting forces as the SAS, and the ghillie suit itself continued to be widely used.

KILT, TARTANS AND BAGPIPES
Traditional

The tartan kilts and bagpipes of a well-rehearsed Scottish regimental band are more than just an impressive sight and sound on parade days. Even more than the ghillie suit, they are integral to the martial tradition of Scotland. And taken together, they are a very special Scottish invention.

The Great Highland Bagpipe is an instrument of war. It is intended to be played in the open air, often in a massed band accompanied by marching drums. Pipe music signals the approach of a mighty army. It is designed to instil fear and awe in the enemy and foster *esprit de corps*.

The kilt has two functions. In former times, it acted as a versatile, multi-purpose garment: the soldier, hunter or shepherd could sleep in it, lie on it, hide in it and run in it. He could fold or arrange it in ways suitable for rain, shine or snow. After the Union of Scotland and England, the tartan kilt eventually became a badge of honour among Scottish soldiers. It was a uniform that set the Scottish regiments apart from their counterparts in the British Army.

In recent years the kilt has enjoyed a huge resurgence in popularity among civilians. This has been accompanied by a debate about how, when and by whom the kilt was invented. It is widely assumed that the kilt is a traditional Scottish garment invented during medieval, or even ancient, times. However, some historians have argued that the kilt is neither old, nor Scottish. Instead, the kilt was dreamed up comparatively recently by an eighteenth-century English industrialist called Thomas Rawlinson.

The story goes that Rawlinson, who in the 1720s owned a charcoal furnace in Glengarry in the Highlands, wanted a suitable working garment for his furnace workers. As a result, he devised what is now known as the kilt. It was only later that the Scots, including the regimental pipe bands, decided to adopt the kilt. In order to explain away the kilt's origins, the Scots concocted a myth that the kilt was a traditional Scottish garment with a very long history.

This story, which is popular among people who dislike the kilt's associations with Scottish nationalism, contains one important flaw: it isn't true. The kilt was, in fact, worn in Scotland long before Rawlinson was born. It is true that Rawlinson did have a kilt tailored for the workers in his furnace, but this kilt was merely a version of a garment that was by then in wide use in the Highlands.

This is not to suggest that Rawlinson played no part in the development of the kilt. Far from it. He deserves credit for helping to ensure the kilt survived as a part of the modern Scot's wardrobe. Rawlinson showed sensitivity and foresight in allowing his Highland workforce to wear their traditional dress by tailoring an example of it that was safe and practical for use at the furnace.

The modern kilt, of which Rawlinson's was an example, is a shortened and refined version of the great kilt. The origins of the great kilt stretch far back into the history of the medieval Highlands, perhaps to the tenth century. The great kilt, or belted plaid, is known in Gaelic as the *feilidh mhor*.

Like the modern kilt, the great kilt is pleated. But the pleats in the great kilt are formed by the wearer folding the material himself and fixing it in place with his belt, rather than the pleats being stitched in. A late example of the great kilt, very similar in appearance to present-day Highland Dress, can be seen in a portrait by Richard Waitt of the piper to the chief of Clan Grant. The painting was made in 1714, around a decade before Rawlinson's alleged invention.

There are other, much earlier illustrations of the kilt. These include a painting from the 1630s of the chief of Clan Campbell. It clearly show the great kilt being worn. An even earlier written description of the Scots wearing the *feilidh mhor* was made in the 1590s. It comes in an account by the chronicler of a battle between Scots from the Western Isles and a force of Irish soldiers. There are also stone carvings in Scotland, made in the earlier 1500s, showing the great kilt being worn.

Is there any evidence that the kilt was in use earlier still? Not much, but equally there is no evidence to prove that it wasn't. Until something new is turned up, the globally famous

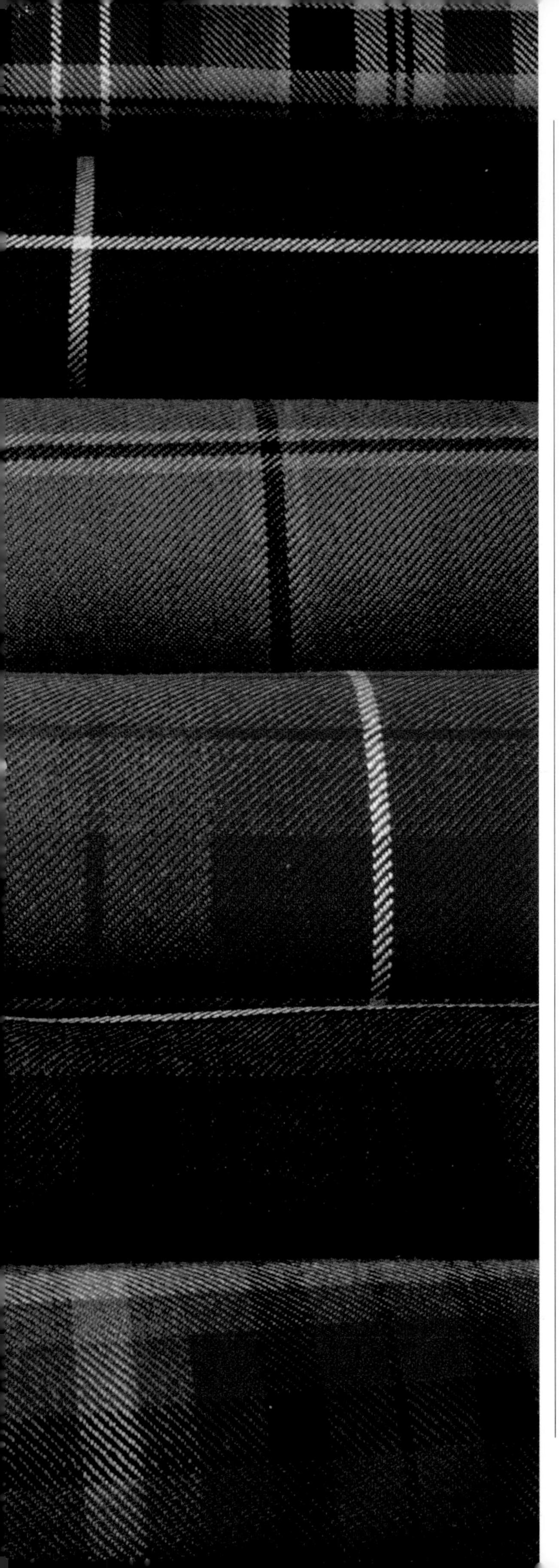

portrayal of kilted thirteenth-century Scots in the 1995 William Wallace biopic *Braveheart* will continue to be derided by some people as bogus. All we can say with certainty is that the kilt is Scottish, it is very old and, in former times, yes, true Scotsmen did not wear any lycra trunks, cotton boxer shorts or terry-towelling Y-fronts under it.

No less contentious is the origin of the bagpipe. It may or may not be the case that the bagpipe was invented in Scotland, although this is unlikely. The instrument's origins are too obscure to make any airtight judgements.

Primitive pipes, albeit bagless, are almost as old as humanity itself. Prehistoric pipe-players used holes cut in the hollow bones of birds and small animals to create simple tunes and rhythms. Certainly this early pipe-playing existed in Scotland, as it did in other parts of the world.

A slightly more recent historical development is the addition of an animal-skin bag and a set of wooden drones to accompany the main pipe 'chanter' on which the melody is played. There are references to what appear to be bagpipes of a sort in classical Greek and other ancient texts. And it is not outwith the realms of possibility that the bagpipe was passed on to the Caledonians of ancient Scotland by the instrumentalists of the Roman Empire.

Later on, the picture becomes clearer. We know that medieval Scotland was among the earliest nations where playing the bagpipe was widespread. It was surely one of the nations that took this instrument closest to heart. In a Scottish poem published by William Dunbar in 1508, the poet writes of 'a bag pipe to play'. Rosslyn Chapel, built near Edinburgh in the fifteenth century, features a carving of an angel playing the bagpipes. However, bagpipes were also popular across Europe, Asia and North Africa, with numerous local variations from Sweden to Algeria.

So much for the bagpipe *per se*. But what of the Great Highland Bagpipe? This distinctive, three-droned set of bagpipes has become the foremost example of the instrument throughout the world. This is thanks largely to its adoption by the Scottish military bands of the former British Empire.

Here the claim for an outright Scottish invention is on much safer ground. The Great Highland Bagpipe is known to have emerged during the late-medieval period in northwest Scotland. By the 1600s, it had become a prestigious and well-known instrument whose leading players, such as the MacCrimmon family on the Isle of Skye, hereditary pipers to the chiefs of the Clan MacLeod, were revered.

When combined with tartan, the distinctive Scottish weave that kilts are usually made from, the kilt and bagpipes form a uniquely Scottish triptych. And like the other two, tartan has contentious origins. The practice of attributing specific tartan patterns to specific clan or family names is rightly regarded as a relatively recent invention, of the eighteenth and nineteenth centuries. Having said that, *some* clans did have their own tartans in earlier times, before it became fashionable for everyone to get in on the act.

As for the tartan check itself, this is reckoned to be a very old Scottish invention indeed. There is a fragment of two-tone tartan cloth now kept in the National Museum in

Edinburgh which is believed to date from around the third century AD. It was found buried inside the Antonine Wall, a turf barrier built between the Forth and Clyde rivers by the soldiers of the Roman Empire during their failed attempt to conquer Scotland.

This cloth, known as the Falkirk Tartan, was made from the wool of the goat-like sheep that then grazed in Scotland. It was probably manufactured using a process that required the cloth to be soaked in urine for several weeks and dyed with peat muck or berry juice.

It is all a far cry from today's electronic bagpipes and trendy kilts. The latter can be made from any number of fabrics, including denim, leather, imitation leopardskin or even rubber. Among aficionados, such kilts are known as MUGs: 'male unbifurcated garments'. One thing we can perhaps all agree on is that William Wallace surely wouldn't have been seen dead in one of those.

CARRONADE CANNON
Robert Melville, 1723–1809

During the revolutionary and Napoleonic Wars of the late 1700s and early 1800s, kilted Scottish regiments made their mark. But an even bigger mark was left by another Scottish invention during the era's famous sea battles.

This was a time when ships went head-to-head in a blaze of cannon fire, smoke and wood splinters. Get your tactics and choice of weapon right and the enemy should be yours for the taking, either sunk outright or else paralysed and ready for boarding. One of the most effective – therefore deadly – pieces of equipment during that historic era proved to be the Carronade cannon.

The Carronade was used by the British Royal Navy. It was a stubby, short-range gun that used less gunpowder than a traditional long-range gun and fired a relatively slow-moving cannonball. But when

Above. A Carronade cannon on HMS *Victory*.

Below. A wall of Carronade cannon on HMS *Victory*.

The Carronade cannon came to be known as 'the Devil's gun'.

used properly it had an unrivalled capacity to wreak utter devastation on an enemy ship. Provided you got close enough – and close-range fighting was the preferred tactic of the Royal Navy – you could smash through the enemy's hull or wipe the crew off their main deck.

First manufactured by the Carron Iron Works in Stirlingshire, the Carronade was invented in 1759 by Robert Melville, a British officer from a well-heeled Fifeshire family. Melville's parents both died when he was young, so he was brought up by his uncle, a professor of medicine at Edinburgh University. Melville was well educated, but tore himself away from a prospective career in medicine to join the armed forces. In his early years he took part in the Jacobite wars and probably participated in the Battle of Culloden, near Inverness, in 1746.

Like Napoleon after him, Melville was fascinated by artillery. He spent a lot of time studying ordnance, which means the development and testing of cannons, ammunition and the like. He came up with the Carronade – which he nicknamed 'the smasher' – to suit the British style of close-up naval fighting. The Carronade, which eventually ranged in calibre from small two-pounders up to huge sixty-eight-pounders, was mounted on a carriage and elevated up or down by a screw. It took a long while for Melville's weapon to catch on, but it impressed Admiral Sir Charles Middleton, who, in 1779, had the Carronade adopted by the Royal Navy.

The Carronade came to be known as 'the Devil's gun' among British seamen. It was one of the most important military inventions of the eighteenth century. By 1781, more than six hundred Carronades were mounted on Royal Navy warships, and the gun proved decisive in several British victories over their enemies, in particular the French, who were outflanked by this remarkable gun for around twenty years until they came up with their own version.

THE UNITED STATES NAVY

John Paul Jones, 1747–1792 with George Washington

Whereas once Britannia ruled the waves, today the oceans have a new master. With almost half a million personnel, almost three hundred ships, close to four thousand aircraft and eleven aircraft carriers, the United States Navy is the largest and most powerful maritime armed force the world has ever known.

The US Navy has come a long way. Its forerunner, known as the Continental Navy, was established in 1775, during America's War of Independence against Britain. It relied on a handful of ships and a deep reserve of sheer courage to take on what was then the world's ultimate naval power, the Royal Navy.

Strictly speaking, the Continental Navy and therefore the US Navy was founded by George Washington. However, at that time it had nothing of the standing that it later enjoyed. From the start, it was under constant threat of being disbanded and forgotten. Certainly, it proved feeble when set against the might of the British Admiralty, as the Royal Navy was then called.

In the early years it was left to one man to galvanise the American Navy as a credible fighting force, thereby ensuring its continued existence. This man was the first person to raise the US ensign over a naval vessel, which he did in 1776. His name was John Paul Jones, then first lieutenant of the frigate USS *Alfred*.

Jones was a gardener's son from Kirkcudbrightshire in Scotland. He went to sea as a teenager, and quickly proved himself an outstanding mariner. For reasons that are unclear, he emigrated to America in his early twenties. He volunteered to join the revolutionaries fighting to free America from British rule. Through skill and good contacts, Jones was able to get his foot in the door among the officers of the Continental Navy. He quickly became a hero, with a string of spectacular naval successes in which he took on and defeated the British in their own coastal waters.

It was during his command of the USS *Bonhomme Richard* that Jones won the new Continental Navy's most spectacular victory. Off the coast of Flamborough Head, east Yorkshire, Jones and his crew engaged the British vessel HMS *Serapis* in combat. During the fight the *Bonhomme Richard* was badly damaged, on fire and taking on water. But when the commander of the *Serapis* asked Jones if he was ready to surrender, Jones simply replied, 'I have not yet begun to fight!'

With the help of another US warship, Jones was able to lash the *Bonhomme Richard* and the *Serapis* together, eventually boarding the *Serapis* and forcing the British commander to surrender. Although the *Bonhomme Richard* later sank, the *Serapis* was commandeered by Jones. He sailed his prize to Holland, a state friendly to the Americans, for repairs.

At a stroke, Jones gave the American revolutionaries a massive morale boost and shattered the legend of British naval superiority. He went on to acquire a fearsome reputation for discipline, professional standards and training – all attributes which he instilled in the early US Navy.

Today Jones is widely regarded as 'the father of the US Navy'. He is interred in a grand bronze and marble sarcophagus at the United States Naval Academy Chapel in Annapolis. And in a declaration often quoted at US naval ceremonies, the spirit of Jones lives on: 'I wish to have no connection with any ship that does not sail fast, for I intend to go in harm's way!'

John Paul Jones.

USS *Bonhomme Richard* and HMS *Serapis* fight it out.

BREECH-LOADING RIFLE
Patrick Ferguson,
1744–1780

Another Scot who eagerly put himself in harm's way during the American War of Independence was Patrick Ferguson. His invention concerns rifles, which, in the 1700s, were used widely on the colonial battlefield. They proved to be much superior to such traditional weapons as swords and spears, yet rifles had some deadly serious limitations.

As a rule, rifles were muzzle-loaded. This meant that the rifleman had to insert his bullet and powder by pushing it down the end of the barrel using a rod. To do this, he usually had to stand or sit up, leaving himself exposed to enemy fire, and

A breech-loading rifle.

the danger was exacerbated by the fact that it took a long time to reload a rifle in this way.

What was needed was a breech-loading rifle. This was a gun into which the rifleman could load his bullet and powder into a 'breech' at the rear of the barrel. It was a much quicker and safer way of doing things. But who could invent such a weapon?

Enter Ferguson, a Scottish army officer from a well-off landed family, the Fergusons of Pitfour in Aberdeenshire. Ferguson served in Germany and the West Indies, where, it is said, he began experimenting with designs for a breech-loading rifle. The breech was opened by the turn of a handle, allowing the infantryman to insert his bullet and powder. Another turn of the handle closed the breech and pushed excess gunpowder into the flashpan. This primed the rifle and made it ready to fire.

Ferguson submitted a patent for his rifle to the British Board of Ordnance in 1776, the same year that the American War of Independence broke out. This presented an opportunity to test the new weapon in the field. Ferguson was given command of a unit of riflemen, all exclusively equipped with the new gun, whom he trained in light-infantry tactics and use of the breech-loading rifle. Sailing from Portsmouth, Ferguson and his men arrived in New York in May 1777.

A month later, Ferguson and his men were in the thick of battle. They took part in the Battle of Brandywine and the British occupation of Philadelphia. Ferguson was commended for his role, and although only twenty-eight of his breech-loading rifles were used in combat, they proved effective.

However, such was the military conservatism of the time that Ferguson's rifle was retired shortly afterwards in favour of more conventional muzzle-loaded rifles. Things probably would have been different if Ferguson hadn't been seriously wounded in combat and rendered unable to move his arms, and therefore unable to continue training men in the use of his rifle.

Some historians have pointed out that Ferguson's design was not the first breech-loading gun. It was, however, the first functional breech-loading rifle, with an accurate, rifled barrel. Moreover, there is not much evidence to suggest that earlier breech-loading designs ever proved their worth in battle. Ferguson's gun did that. It has also been claimed that the mechanism in Ferguson's gun became too easily fouled up by expended gunpowder, which caused it to lose accuracy or seize up entirely. Yet there is no real evidence to support the claim that the gun was an abject failure.

Despite its limitations, Ferguson's weapon did make a real contribution to the British victory at Brandywine. In doing so, it paved the way for the adoption of breech-loading rifles as standard in the mid-1800s, with the invention of the self-contained metallic ammunition cartridge.

PERCUSSION CAP
Alexander John Forsyth, 1769–1843

If you were ever guilty of scaring the wits out of unsuspecting grown-ups by firing a toy pistol at them, then you will be familiar with the basic concept of the percussion cap. The 'caps' that come on the roll of paper ammunition for a toy gun are ignited simply by being struck by the gun's hammer, thanks to the special properties of the materials used in the cap. Result: BANG!

The percussion cap was invented back in the 1800s, not as a toy, but as an ignition system for real guns. It was the foreruner of the self-contained breech-loading metal cartridge, which is now the typical ammunition of guns everywhere, and which has an integrated percussion cap.

The inventor of the percussion cap was Alexander John Forsyth, a Church of Scotland minister from Belhelvie in Aberdeenshire. He came up with the idea in 1805 while indulging in one of his favourite pastimes, hunting wild birds. Forsyth noticed that a traditional flintlock rifle gave out a puff of smoke during ignition that acted as an early warning system for the birds, giving them time to dive out of the way before the bullet was released.

To try to come up with a better system, Forsyth began experimenting with detonating compounds. Eventually he found a substance called fulminating powder. He saw that fulminating powder acted as a powerful ignition charge for a gun when it was struck by a hammer.

Forsyth constructed a gun with a percussion cap using a metal tube containing fulminating powder. When the end of the tube was struck by the gun's trigger hammer, the fulminating powder ignited. This sent a spark out of the other end of the tube, igniting the main charge of gunpowder and firing the bullet. When Forsyth tested the gun he found it was a great improvement: there was no puff of smoke and the time between the pulling of the trigger and the shot being released was reduced.

Early in 1806, Forsyth demonstrated his new gun in London. News of his invention reached Lord Moira, Master-General of the Ordnance, who ordered that the British armed forces should make use of the new gun. The government initially prevented Forsyth from patenting his invention, but this ruling was later reversed. In 1808 Forsyth set up the Forsyth Patent Gun Company with his cousin, James Brougham, to make percussion-cap firearms. The name was later changed to Forsyth & Co. Patent Gunmakers.

In 1843 the government awarded Forsyth £1000 as thanks for the use of his invention by the British armed forces. But he died before the award could be made, and the money was divided among his surviving relatives.

In later years, refinements were made to Forsyth's invention to make it easier to use in the heat of battle, especially while on horseback. One such example was the Maynard Tape Primer, which used a roll of paper caps similar to those used by a toy gun. This system was adopted for some firearms, such as the Springfield .58 rifle musket, during the American Civil War.

Percussion caps.

CORDITE
James Dewar, 1847–1932 with Frederick Augustus Abel

The whiff of cordite is a well-kent description of battle. But what is cordite and, more importantly, who invented it? Cordite is a smokeless propellant used in gun barrels to launch bullets or shells. It is deliberately designed as a low-explosive, so that when ignited in a tank barrel, for example, it will explode with enough force to propel the tank shell to its target – but without destroying the tank's gun barrel in the process.

Cordite is now obsolete, but it was used widely on the battlefield until the late twentieth century. It came in spaghetti-like rods, which were made from a compound of different substances. It was invented in the 1890s by James Dewar, a chemist from Kincardine-on-Forth. As a child, Dewar had demonstrated his ingenuity and handicraft skills by making fiddles while recuperating from rheumatic fever. His cordite co-inventor was Frederick Augustus Abel, a well-heeled Anglo-German explosives expert.

Dewar and Abel devised a compound of nitroglycerine, nitrocellulose (or guncotton) and vaseline, using acetone as a solvent to combine all the ingredients. Their invention was patented in 1889, only to be challenged by Alfred Nobel, the inventor of dynamite, who claimed that Dewar and Abel's design was copied from his earlier patent called Ballistite. However, the authorities disagreed, and Dewar and Abel won the right to be remembered as the inventors of cordite.

GAS MASK
John Stenhouse, 1809–1880

Like a loaded gun, the gas mask can be a scary-looking piece of kit. It makes people appear like aliens and prompts many of us to think of wartime air-raids and deadly chemical weapons.

However, while the gas mask is often used in combat situations by troops and civilians alike, it was not originally invented for war. Nor was the gas mask the work of any one individual alone. During the nineteenth century, several people had the idea of an apparatus that would protect people from breathing in harmful substances.

One of the earliest and best designs for a gas mask was invented in the 1850s by John Stenhouse, a scientist from Glasgow who learned his profession at Glasgow University from some of the finest chemists then working. Stenhouse was a classic inventor. He held patents for dyes, methods of waterproofing, tanning and making sugar. His mask arose from his investigations into the absorptive power of charcoal.

Chemists had long known that charcoal could remove unpleasant odours from the air by absorbing them. But Stenhouse revealed that it also had the capacity to remove such potentially harmful chemicals as ammonia, chlorine and hydrogen sulphide. He designed a prototype protective breathing mask, which contained powdered wood charcoal, sandwiched between two hemispheres of wire gauze. It was tightly moulded to the wearer's face using velvet-lined lead edging, and held in place by elastic bands.

In 1854, Stenhouse demonstrated his gas mask to the

Society of Arts. He did not patent his invention, but offered it to the public. It was soon realised that the mask could protect chemical workers, for example, from exposure to toxic vapours in factories. It was also realised that it could protect soldiers from chemical-weapons attack, and to that end an example was examined by the Board of Ordnance. Using the same technology, Stenhouse also developed an air filter for use in public buildings. He was later awarded a medal by the Royal Society for his work.

RADAR
Robert Watson-Watt, 1892–1973

Like the gas mask, Radar is an icon of wartime. Unlike the gas mask, it was invented with war in mind. It first appeared in the 1930s, on the eve of the Second World War.

At that time Britain was growing anxious about reports that Hitler's Germany was developing an electromagnetic 'death ray' as a potentially devastating secret weapon. So the British set about trying to see whether they could develop a death ray of their own. The aim of the British death ray was to fire a beam of energy at a German bomber, causing the plane to heat up. Conditions on board would become intolerable for the crew. In the ultimate scenario, the ray would cause the plane to melt.

However, the death-ray concept was soon found to be more science fiction than science fact. A study conducted on behalf of the British Air Ministry concluded that it was not practical, and that the Germans must have realised the same. But the study was not a complete waste of time. During it, the British scientists made an important discovery. That discovery was presented to the Air Ministry in 1935 in a proposal called *The Detection of Aircraft by Radio Methods*. In other words, a proposal for Radar.

The leader of the team which drafted the Radar proposal was Robert Watson-Watt, a Scottish scientist from Brechin, Forfarshire. Watson-Watt and his team had discovered that, although a beam of electromagnetic energy could not be used to destroy an aircraft, it could be used to detect where it was, how far away it was, and how fast it was moving. This kind of detection would be vital when it came to identifying German bombers early enough to shoot them down before they had the chance to make an air-raid on British cities.

Radar, which is short for 'Radio detection and ranging', sounds complicated. But its basic principles are simple. First, a dish transmits pulses of electromagnetic energy, such as radio waves or microwaves, into the air. These pulses bounce off any object, such as an aircraft, in their path. When the pulses bounce off the object, a small amount of energy is reflected back to where it has come from.

This reflected energy is then picked up by another dish, or antenna. Scientists then calculate how long it took for the waves to be reflected. In doing so, they can work out how far away the object is, how fast it is moving, and other important information. In a classic Radar system, the information shows up in a circular image on a TV screen, which is refreshed over and over again as the Radar dish sends out fresh pulses of energy.

It didn't take long for Watson-Watt's system to be taken from the drawing board and turned into reality. After all, Watson-Watt – whose interest in electromagnetic energy was first stimulated during his education at St Andrews University College, Dundee – was trusted by the British authorities. In earlier years, he had shown how radio waves could be used to detect thunderstorms. So when he demonstrated a prototype of the Radar system to his superiors, they agreed to build a chain of Radar stations to defend Britain in case of German air attack.

It was a wise move. By the time Watson-Watt and his team had set up a reliable network of coastal Radar stations in 1940, the Second World War had been declared and Germany had begun attacking Britain from the air. It has often been said that if it wasn't for Watson-Watt's Radar system detecting German bombers in time for British fighters and ground-to-air batteries to intercept and destroy them, then Britain might have lost the war. In the years after the war, while Watson-Watt received many honours for his great achievement, Radar came to be a vital part of the world's transport and communications network.

By the time of its inventor's death in Inverness in 1973, Radar had become an essential navigational tool on board passenger aircraft and shipping. Employed by weather stations to predict storms, Radar is also widely used today for monitoring traffic speed. It is a fine example of what was originally a wartime invention being used to promote a safe, peaceful and prosperous way of life.

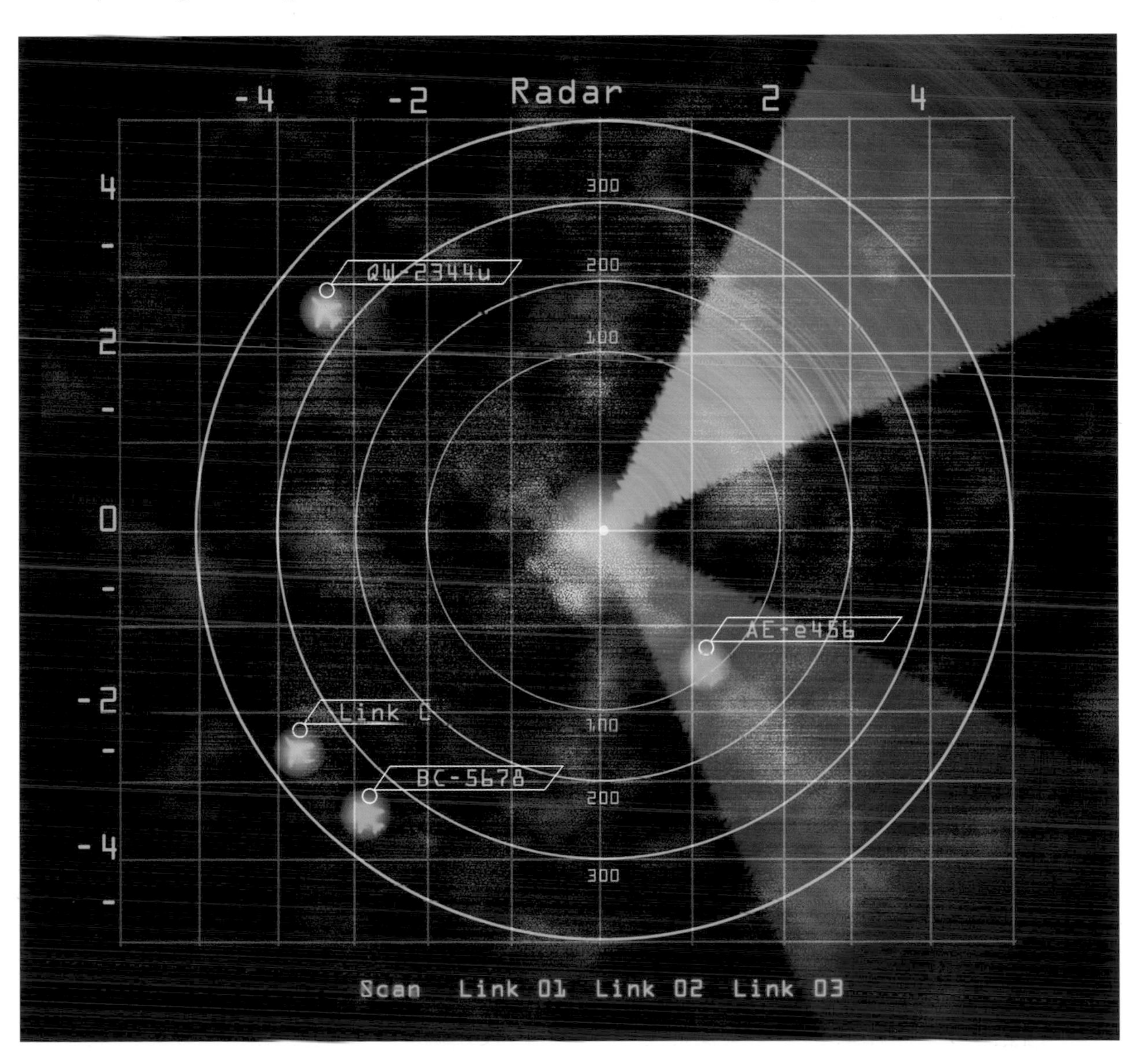

SELECTED BIBLIOGRAPHY AND FURTHER READING

Various authors, *The Oxford Dictionary of National Biography* (2004 Edition; 2009 Supplement)

Various authors, *The Encyclopaedia Britannica* (2009 Digital Edition)

Elspeth Wills, *Scottish Firsts: A Celebration of Innovation and Achievement* (2002)

Carol Foreman, *Made in Scotland* (2004)

Christopher Winn, *I Never Knew That about Scotland* (2007)

Bill Fletcher, *Great Scottish Discoveries and Inventions* (1985)

Richard Platt, *Eureka! Great Inventions and How They Happened* (2003)

Russell Burns, *John Logie Baird: Television Pioneer* (2001)

Robert Bruce, *Alexander Graham Bell and the Conquest of Solitude* (1990)

Ian Simpson Ross, *The Life of Adam Smith* (1995)

Luetta Reimer and Wilbert Reimer, *Mathematicians Are People Too: Stories from the Lives of Great Mathematicians* (1992)

Hugh Cheape, *Tartan: The Highland Habit* (2006)

Hugh Cheape, *Bagpipes: A National Collection of a National Treasure* (2008)

Alan Gauld, *A History of Hypnotism* (1992)

Myrtle Simpson, *Simpson The Obstetrician* (1972)

John Griffiths, *The Third Man: The Life and Times of William Murdoch* (1992)

Dennis Dean, *John Muir and the Origins of Yosemite Valley, Annals of Science* Vol 48 (1991)

James Mackay, *Allan Pinkerton: The Spy Who Never Slept* (1996)

John Strawson, *A History of the SAS Regiment* (1984)

James Bradford, *John Paul Jones and the American Navy* (2002)

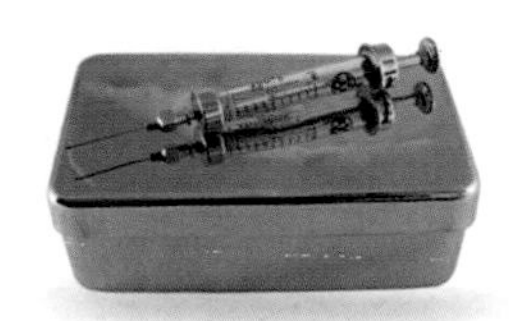

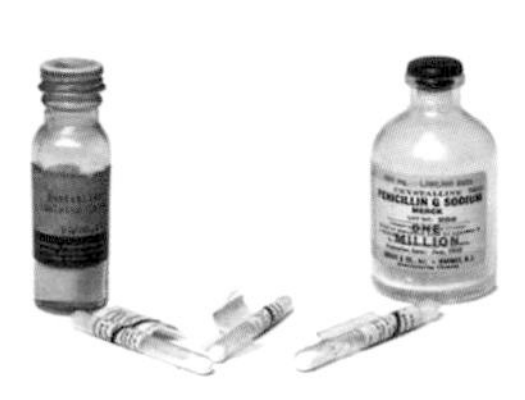